HEATING SYSTEMS TROUBLESHOOTING HANDBOOK

Library of Congress Cataloging-in-Publication Data

Langley, Billy C.
    Heating systems troubleshooting handbook.

    "A Reston book."
    Includes index.
    1. Heating—Equipment and supplies—Maintenance
and repair. I. Title.
TH7225.L36 1988   697′.07   87-14517
ISBN 0-8359-2805-5

Editorial/production supervision and
    interior design: Eileen M. O'Sullivan
Cover design: 20/20 Services, Inc.
Manufacturing buyer: Lorraine Fumoso

Printed in the United States of America

10 9 8 7 6 5 4 3 2 1

ISBN 0-8359-2805-5 025

PRENTICE-HALL INTERNATIONAL (UK) LIMITED, *London*
PRENTICE-HALL OF AUSTRALIA PTY. LIMITED, *Sydney*
PRENTICE-HALL CANADA INC., *Toronto*
PRENTICE-HALL HISPANOAMERICANA, S.A., *Mexico*
PRENTICE-HALL OF INDIA PRIVATE LIMITED, *New Delhi*
PRENTICE-HALL OF JAPAN, INC., *Tokyo*
SIMON & SCHUSTER ASIA PTE. LTD., *Singapore*
EDITORA PRENTICE-HALL DO BRASIL, LTDA., *Rio de Janeiro*

Contents

Preface

HOW TO USE THIS HANDBOOK

When trouble is experienced with a particular type of equipment, turn to the chart for that type of unit in Chapter 1, Troubleshooting Charts. The action of the unit can be found in the column labeled "condition." The next column has the "possible causes" that may cause the condition encountered. The third column lists the "corrective action" that may be taken to correct the problem. The "reference" column directs the reader to the specific sections in Chapter 2 where the component is introduced and where checkout procedures for the component is given. Specific procedures described in Chapter 4, Standard Service Procedures, will help beginners to accomplish any tasks that may be unfamiliar.

Example: We are having trouble with a heat pump unit in the heating cycle. After checking the unit we find that the compressor runs but there is no heating. When we refer to the troubleshooting chart, "Heat Pump (Heating Cycle)" the first cause indicated is refrigerant shortage. The corrective action is to repair the leak and recharge the system. The reference column indicates the specific sections in the text where the proper procedure for this job is discussed. These specific sections are Chapter 2, Sections 2-26 and 2-26A. A check in Chapter 4 will provide additional information on making these repairs.

Also included in this handbook are startup procedures for heating systems and heat pump systems (Chapter 3), standard service procedures and wiring diagrams (Chapter 4), flame types and combustion troubleshooting (Chapter 5), safety procedures (Chapter 6), and useful engineering data (Chapter 7).

Billy C. Langley

ACKNOWLEDGMENTS

Material in Chapters 2, 3, 4, 5, and 6 was originally published in the following titles:

Billy C. Langley, AIR CONDITIONING AND REFRIGERATION TROUBLE-SHOOTING HANDBOOK, © 1980 chapters 2, 3, 4, and 6. (A Reston Publication.) Reprinted by permission of Prentice Hall, Englewood Cliffs, New Jersey, 07632.

Billy C. Langley, COOLING SYSTEMS TROUBLESHOOTING HANDBOOK, © 1986 chapter 2. (A Reston Publication.) Reprinted by permission of Prentice Hall, Englewood Cliffs, New Jersey, 07632.

Billy C. Langley, COMFORT HEATING 3E, © 1985 chapters 3 and 13. (A Reston Publication.) Reprinted by permission of Prentice Hall, Englewood Cliffs, New Jersey, 07632.

All illustrations without courtesy lines are taken from the above three books.

The cover illustration is courtesy of White-Rodgers Division, Emerson Electric Co.

The diagrams used in pages 211 to 226 are courtesy of Lennox Industries. The diagrams used in pages 227 to 245 are courtesy of Magic Chef.

HEATING SYSTEMS
TROUBLESHOOTING
HANDBOOK

1 Troubleshooting Charts

1-1 ELECTRIC HEAT

Condition	Possible cause	Corrective action	Reference
Unit will not run.	1. Blown fuse.	1. Replace fuse and correct cause.	2-3
	2. Burned transformer.	2. Replace transformer and correct cause.	2-36;2-36A
	3. Thermostat not calling for heat.	3. Set thermostat.	2-13;2-13A,B,C,D
	4. Defective thermostat.	4. Replace thermostat.	2-13;2-13A,B,C,D
	5. Defective heating relay.	5. Replace relay.	2-43
Fan will not run.	1. Burned fan motor.	1. Repair or replace fan motor.	2-35;2-35A, B,C,D,E,F
	2. Broken fan belt.	2. Replace fan belt.	2-26E
	3. Burned contacts in fan relay.	3. Replace fan relay.	2-39
	4. Defective fan control.	4. Replace fan control.	2-44;2-44A,B
	5. Defective wiring or connections.	5. Repair wiring or connections.	2-15
Fan motor hums but will not start.	1. Defective fan motor bearings.	1. Replace bearings or fan motor.	2-35;2-35D
	2. Defective fan motor starting switch.	2. Repair starting switch or replace motor.	2-35;2-35D,F

(continued)

1

1-1 ELECTRIC HEAT (continued)

Condition	Possible cause	Corrective action	Reference
	3. Defective starting capacitor.	3. Replace capacitor.	2-18;2-18A
	4. Burned start winding in motor.	4. Repair or replace motor.	2-35;2-35A, B,C,D,E,F
	5. Defective blower bearings.	5. Replace bearings.	2-48;2-48B
	6. Loose wiring connections in motor starting circuit.	6. Repair wiring.	2-15
Fan motor cycles.	1. Defective fan motor bearings.	1. Replace bearings or motor.	2-35;2-35E
	2. Defective blower bearings.	2. Replace blower bearings.	2-48;2-48B
	3. Defective run capacitor.	3. Replace capacitor.	2-18;2-18B
	4. Defective fan control.	4. Replace fan control.	2-44;2-44A,B
	5. Defective fan relay.	5. Replace relay.	2-39
	6. Defective motor windings.	6. Repair or replace motor.	2-35;2-35A,B,C,F
Fan blows cold air.	1. Defective heat sequencing relays.	1. Replace relays.	2-45;2-46A,B,C
	2. Burned heat elements.	2. Replace elements.	2-46;2-46A,B,C
	3. Loose wiring connections.	3. Repair wiring.	2-15
	4. Defective thermostat.	4. Replace thermostat.	2-13;2-13A,B,C,D
	5. Fan set to "on" position.	5. Set to "auto" position.	2-13;2-13D
	6. Defective fan control.	6. Replace fan control.	2-44;2-44A,B
Not enough heat.	1. Dirty air filters.	1. Clean or replace filters.	2-26C
	2. Unit too small.	2. Install more elements.	2-41;2-46
	3. Too little airflow through furnace.	3. Increase airflow; remove restrictions.	2-47
	4. Thermostat heat anticipator not properly set.	4. Set heat anticipator.	2-13;2-13B
	5. Defective fan motor.	5. Repair or replace fan motor.	2-35;2-35A, B,C,D,E,F
	6. Air-conditioning evaporator dirty.	6. Clean evaporator.	2-26;2-26C
	7. Thermostat not properly located.	7. Relocate thermostat.	2-13;2-13C
	8. Thermostat set too low.	8. Set thermostat.	2-13;2-13A

1–1 ELECTRIC HEAT (continued)

Condition	Possible cause	Corrective action	Reference
	9. Thermostat out of calibration.	9. Calibrate thermostat.	2-13;2-13A
	10. Low voltage.	10. Correct cause.	2-17
	11. Air ducts not insulated.	11. Insulate ducts.	2-40
	12. Burned elements.	12. Replace elements.	2-46;2-46A,B,C
	13. Defective heat sequencing relays.	13. Replace relays.	2-45;2-45A,B,C
	14. Defective thermostat.	14. Replace thermostat.	2-13;2-13A, B,C,D,E
	15. Defective element limits.	15. Replace limits.	2-46;2-46A,B,C
	16. Outdoor thermostat set too low.	16. Reset thermostat.	2-13;2-13E
Too much heat.	1. Unit too large.	1. Reduce Btu input.	2-49
	2. Thermostat heat anticipator not properly set.	2. Set heat anticipator.	2-13;2-13B
	3. Thermostat not properly located.	3. Relocate thermostat.	2-13;2-13C
	4. Thermostat set too high.	4. Set thermostat.	2-13;2-13A
	5. Thermostat out of calibration.	5. Calibrate thermostat.	2-13;2-13A,B
High humidity in building.	1. Humidity due to cooking.	1. Vent cookstove.	2-50;2-57
	2. Humidity due to bathing.	2. Vent bathroom.	2-50
	3. Humidity due to rain.	3. Increase temperature rise through furnace.	2-47;2-50
Blown element limits.	1. Shorted heating element.	1. Replace element and correct cause.	2-46;2-46A,B,C
	2. Dirty filters.	2. Clean or replace filters.	2-26B
	3. Dirty blower.	3. Clean blower.	2-26D
	4. Broken or slipping fan belt.	4. Adjust or replace belt.	2-26E
	5. High or low voltage.	5. Notify power company.	2-1;2-17
	6. Defective blower motor.	6. Replace motor.	2-35;2-35A, B,C,D,E,F
	7. Not enough air through furnace.	7. Remove restriction or increase blower speed.	2-26;2-26B,C,D,E; 2-35A,B,C,D,E,F
	8. Loose electrical connections.	8. Repair connections.	2-15

(continued)

1–1 ELECTRIC HEAT (continued)

Condition	Possible cause	Corrective action	Reference
High operating costs.	1. Unit too small.	1. Increase number of elements.	2-41;2-46A,B,C
	2. Dirty air filters.	2. Clean or replace filters.	2-26B
	3. Dirty air-conditioning evaporator.	3. Clean evaporator.	2-26C
	4. Air ducts not insulated.	4. Insulate ducts.	2-40
	5. Thermostat in wrong location.	5. Relocate thermostat.	2-13;2-13C
	6. Dirty blower.	6. Clean blower.	2-26D
	7. Defective thermostat.	7. Replace thermostat.	2-13;2-13A,B,C,D
	8. Fan belt slipping.	8. Replace or adjust belt.	2-26E
	9. Low or high voltage.	9. Notify power company.	2-1;2-17
	10. Thermostat setting too high.	10. Set to lower setting.	2-13;2-13A

1–2 GAS HEAT

Condition	Possible cause	Corrective action	Reference
Unit will not run.	1. Blown fuse.	1. Replace fuse and correct cause.	2-3
	2. Burned transformer.	2. Replace transformer and correct cause.	2-36;2-36A
	3. Pilot out.	3. See "Pilot not burning properly or out" (p. 5).	
	4. Thermostat not calling for heat.	4. Set thermostat.	2-13;2-13A
	5. Defective wiring or connections.	5. Repair wiring or connections.	2-15
Fan will not run.	1. Burned fan motor.	1. Repair or replace motor.	2-35;2-35A, B,C,D,E,F
	2. Broken fan belt.	2. Replace fan belt.	2-26E
	3. Burned contacts in fan relay.	3. Replace fan relay.	2-39
	4. Defective fan control.	4. Replace fan control.	2-44;2-44A,B
	5. Defective wiring or connections.	5. Repair wiring or connections.	2-15

1–2 GAS HEAT (continued)

Condition	Possible cause	Corrective action	Reference
Fan motor hums but will not start.	1. Defective bearings in fan motor.	1. Replace bearings or fan motor.	2-35;2-35A, B,C,D,E,F
	2. Defective starting switch in fan motor.	2. Replace starting switch or replace motor.	2-35;2-35D
	3. Defective starting capacitor.	3. Replace capacitor.	2-18;2-18A,B
	4. Burned start winding in motor.	4. Repair or replace motor.	2-35;2-35A, B,C,D,E,F
	5. Defective blower bearings.	5. Replace bearings.	2-48;2-48A,B
	6. Loose wiring connections in motor starting circuit.	6. Repair wiring.	2-15
Fan motor cycles.	1. Defective motor bearings.	1. Replace bearings or fan motor.	2-35;2-35E,F
	2. Defective blower bearings.	2. Replace blower bearings.	2-48;2-48B
	3. Defective run capacitor.	3. Replace run capacitor.	2-18;2-18B
	4. Defective fan control.	4. Replace fan control.	2-44;2-44A,B
	5. Return air too cool.	5. Allow air to warm.	2-44;2-44B
	6. Fan control differential too close.	6. Adjust fan control.	2-44;2-44A,B
	7. Fan control "off" setting too high.	7. Adjust fan control.	2-44;2-44A,B
	8. Too much airflow through furnace.	8. Reduce airflow.	2-44;2-44A,B
	9. Defective motor windings.	9. Repair or replace motor.	2-35;2-35A,B,C,F
	10. Fan control "on" setting too low.	10. Adjust fan control.	2-44;2-44A,B
Pilot not burning properly or out.	1. Faulty thermocouple.	1. Replace thermocouple.	2-51;2-51A
	2. Dirty or corroded thermocouple connection.	2. Clean connection.	2-51;2-51A
	3. Gas supply turned off.	3. Restore gas supply.	
	4. Pilot burner orifice dirty.	4. Clean orifice.	2-51B;2-52A,B,C
	5. Thermocouple not installed in flame properly.	5. Properly install thermocouple.	2-51;2-51A
	6. Drafts affecting pilot flame.	6. Shield pilot from drafts.	2-51;2-51B

(continued)

1-2 GAS HEAT (continued)

Condition	Possible cause	Corrective action	Reference
	7. Defective pilot safety device.	7. Replace pilot safety device.	2-51;2-51C
Fan cycles while burner stays on.	1. Wrong-size orifices in burners.	1. Replace orifices with proper size.	2-52;2-52A,C
	2. Low manifold gas pressure.	2. Increase gas pressure.	2-55;2-55A,B
	3. Too much air flowing through furnace.	3. Reduce airflow.	2-44;2-44B
	4. Too cold return air.	4. Allow return air to warm.	
	5. Electrical or motor problems.	5. See ''Fan motor cycles'' (p. 5).	
Not enough heat.	1. Dirty air filters.	1. Clean or replace filters.	2-26B
	2. Wrong-size orifices in burners.	2. Replace with orifices of proper size.	2-52;2-52A,C
	3. Low manifold gas pressure.	3. Increase gas pressure.	2-55;2-55A,B
	4. Too little airflow through furnace.	4. Increase airflow; clear restrictions.	2-26B,C,D,E
	5. Thermostat heat anticipator not properly set.	5. Set heat anticipator.	2-13;2-13B
	6. Defective fan motor.	6. Replace or repair fan motor.	2-35;2-35A, B,C,D,E,F
	7. Unit too small.	7. Replace unit with one of proper size.	2-41
	8. Air-conditioning evaporator dirty.	8. Clean evaporator.	2-26C
	9. Thermostat not properly located.	9. Move thermostat.	2-13;2-13C
	10. Thermostat set too low.	10. Set thermostat.	2-13;2-13A
	11. Thermostat out of calibration.	11. Calibrate thermostat.	2-13;2-13A
Too much heat.	1. Unit too large.	1. Reduce Btu input; replace unit.	2-49
	2. Thermostat heat anticipator not properly set.	2. Set heat anticipator.	2-13;2-13B
	3. Thermostat not properly located.	3. Move thermostat.	2-13;2-13C
	4. Thermostat set too high.	4. Set thermostat.	2-13;2-13A,B
	5. Thermostat out of calibration.	5. Calibrate thermostat.	2-13;2-13A,B

1–2 GAS HEAT (continued)

Condition	Possible cause	Corrective action	Reference
Pilot burning; main gas valve will not operate.	1. Blown fuse.	1. Replace fuse and check for cause.	2-3
	2. Defective gas valve.	2. Replace gas valve.	2-54;2-54A,B
	3. Burned transformer.	3. Replace transformer and check for cause.	2-36;2-36A
	4. Burned thermostat heat anticipator.	4. Replace thermostat.	2-13;2-13B
	5. Bad thermostat.	5. Replace thermostat.	2-13;2-13B,D
	6. Bad electrical connections.	6. Repair connections.	2-15
	7. Broken thermostat wire.	7. Repair broken wire.	2-15
Delayed ignition of main burner.	1. Poor flame travel to the burner.	1. Correct flame travel.	2-53;2-53A
	2. Poor flame distribution over the burner.	2. Correct flame distribution.	2-53;2-53B
	3. Low manifold gas pressure.	3. Adjust gas pressure.	2-55;2-55B
	4. Defective step-opening regulator.	4. Adjust or replace regulator.	2-55;2-55A
Roll-out on main burner ignition.	1. Restricted heat exchanger.	1. Clean restrictions.	2-56;2-56A
	2. Quick opening main gas valve.	2. Install surge arrester.	2-54;2-54A
Flame flashback.	1. Low manifold gas pressure.	1. Adjust manifold gas pressure.	2-55;2-55A
	2. Extremely small main burner flame.	2. Adjust primary air.	2-53;2-53C
	3. Distorted burner or carryover wing slots.	3. Repair burner or carryover wing slots.	2-53;2-53A
	4. Defective main burner orifice.	4. Replace orifices.	2-52;2-52A,B,C
	5. Orifice misaligned.	5. Replace orifices.	2-52;2-52B
	6. Erratic gas valve operation.	6. Replace gas valve.	2-54
	7. Dirty burner.	7. Clean burners.	2-53
	8. Improper gas–air mixture.	8. Be sure that proper gas is being used.	2-53;2-53C
	9. Unstable gas supply pressure.	9. Install a two-stage pressure regulator.	2-55;2-55B
Resonance (loud, rumbling noise).	1. Excess primary air to main burner.	1. Adjust primary air.	2-53;2-53C
	2. Defective orifice spud.	2. Replace orifice spud.	2-52;2-52A,B,C
	3. Dirty orifice spuds.	3. Clean orifices.	2-52

(continued)

1-2 GAS HEAT (continued)

Condition	Possible cause	Corrective action	Reference
Yellow flame.	1. Too little primary air.	1. Adjust primary air.	2-53;2-53C
	2. Dirty orifice spud.	2. Clean spuds.	2-52
	3. Orifice spuds misaligned.	3. Align orifice spuds.	2-52;2-52B
	4. Restricted heat exchanger.	4. Clean heat exchanger.	2-56;2-56A
	5. Poor vent operation.	5. Correct venting.	2-57
Floating main burner flame.	1. Air blowing into heat exchanger.	1. Check for defective heat exchanger.	2-56;2-56B
	2. Restricted heat exchanger.	2. Clean heat exchanger.	2-56;2-56A
	3. Negative pressure in furnace room.	3. Increase air supply to room.	2-57
Main burner flame too large.	1. Orifices too large.	1. Replace orifices.	2-52;2-52A
	2. Excessive manifold gas pressure.	2. Adjust pressure regulator.	2-55;2-55A,B
	3. Defective gas pressure regulator.	3. Replace regulator.	2-55
	4. Wrong type of gas being used.	4. Install changeover kit.	2-52
Main burner flame too small.	1. Dirty orifice spuds.	1. Clean orifice spuds.	2-52
	2. Low manifold gas pressure.	2. Adjust pressure regulator.	2-55
	3. Orifices too small.	3. Replace orifices.	2-52;2-52C
	4. Wrong type of gas being used.	4. Install changeover kit.	2-52
Odor in building.	1. Vent not operating properly.	1. Correct venting problem.	2-57
	2. Poor ventilation.	2. Check the flame conditions and correct.	2-57
High operating costs.	1. Unit too small.	1. Install proper-size unit.	2-41
	2. Dirty air filters.	2. Clean or replace filters.	2-26B
	3. Dirty air-conditioning evaporator.	3. Clean evaporator.	2-26C
	4. Air ducts not insulated.	4. Insulate ducts.	2-40
	5. Thermostat in wrong location.	5. Relocate thermostat.	2-13;2-13C
	6. Dirty blower.	6. Clean blower.	2-26D

1-3 HEAT PUMP (HEATING CYCLE)

Condition	Possible cause	Corrective action	Reference
No heating but compressor runs continuously.	1. Refrigerant shortage.	1. Repair leak and recharge.	2-26;2-26A
	2. Compressor valves defective.	2. Replace valves and plate or compressor.	2-7;2-7C
	3. Leaking reversing valve.	3. Replace reversing valve.	2-61
	4. Defective defrost control, time clock, or relay.	4. Replace defrost control, time clock, or relay.	2-62
Too much heat; compressor runs continuously.	1. Faulty wiring.	1. Repair wiring.	2-15
	2. Faulty thermostat.	2. Replace thermostat.	2-13;2-13A
	3. Wrong thermostat location.	3. Relocate thermostat.	2-13;2-13C
Compressor cycles on low-pressure control at end of defrost cycle.	1. Defective reversing valve.	1. Replace reversing valve.	2-61
	2. Defective power element on indoor expansion valve.	2. Replace power element.	2-27;2-27E
	3. Shortage of refrigerant.	3. Repair leak and recharge.	2-26;2-26A
Unit runs in cooling cycle but pumps down in heating cycle.	1. Faulty outdoor expansion valve.	1. Clean or replace expansion valve.	2-27;2-27A, B,C,D,E
	2. Defective power element on outdoor expansion value.	2. Replace power element.	2-27;2-27E
	3. Defective reversing valve.	3. Replace reversing valve.	2-61
	4. Dirty outdoor coil.	4. Clean coil.	2-25;2-25B,C
	5. Belt slipping on outdoor blower.	5. Replace or adjust belt.	2-26E
	6. Defective indoor check valve.	6. Replace check valve.	2-60
	7. Restriction in refrigerant circuit.	7. Locate and remove restriction.	2-28;2-28A,B,C,D
Defrost cycle will not terminate.	1. Shortage of refrigerant.	1. Repair leak and recharge.	2-26;2-26A
	2. Defrost control out of adjustment.	2. Adjust control.	2-62
	3. Defective defrost control, time clock, or relay.	3. Replace defrost control, time clock, or relay.	2-62
	4. Defective reversing valve.	4. Replace reversing valve.	2-61

(continued)

1-3 HEAT PUMP (HEATING CYCLE) (continued)

Condition	Possible cause	Corrective action	Reference
	5. Defective compressor valves.	5. Replace valves and valve plate or compressor.	2-7;2-7C
	6. Faulty electrical wiring.	6. Repair wiring.	2-15
Defrost cycle initiates without ice on coil.	1. Shortage of refrigerant.	1. Repair leak and recharge.	2-26;2-26A
	2. Defrost control out of adjustment.	2. Adjust control.	2-62
	3. Defective defrost control, time clock, or relay.	3. Replace defrost control, time clock, or relay.	2-62
	4. Defrost control sensing element not making good contact.	4. Improve contact.	2-62
	5. Outdoor coil dirty.	5. Clean coil.	2-26;2-26C
	6. Outdoor fan belt slipping.	6. Replace belt or adjust.	2-26;2-26E
Reversing valve will not shift.	1. Defective reversing valve.	1. Replace reversing valve.	2-61
	2. Defective compressor valves.	2. Replace valves and valve plate or compressor.	2-7;2-7C
	3. Faulty fan relay on either indoor or outdoor section.	3. Replace relay.	2-39
	4. Burned transformer.	4. Replace transformer.	2-36;2-36A
Indoor blower off with auxiliary heat on.	1. Defective indoor fan relay.	1. Replace fan relay.	2-39
	2. Defective indoor fan motor.	2. Repair or replace motor.	2-35;2-35A, B,C,D,E,F
	3. Faulty wiring or loose terminals.	3. Repair wiring or terminals.	2-15
	4. Faulty thermostat.	4. Replace thermostat.	2-13;2-13A,B,C,D
Outdoor blower runs during defrost cycle.	1. Faulty outdoor fan relay.	1. Replace relay.	2-42
Compressor short cycles on defrost control.	1. Shortage of refrigerant.	1. Repair leak and recharge.	2-26;2-26A
	2. Defrost control out of adjustment.	2. Adjust defrost control.	2-62
	3. Defective defrost control, time clock, or relay.	3. Replace defrost control, time clock, or relay.	2-62

1-3 HEAT PUMP (HEATING CYCLE) (continued)

Condition	Possible cause	Corrective action	Reference
	4. Defective power element on outdoor expansion valve.	4. Replace power element.	2-27;2-27E
	5. Fan belt slipping on outdoor blower.	5. Replace or adjust belt.	2-26E
Excessive ice buildup on indoor coil.	1. Defective defrost relay.	1. Replace defrost relay.	2-65
	2. Defective compressor valves.	2. Replace valves and valve plate or compressor.	2-7;2-7C
	3. Shortage of refrigerant.	3. Repair leak and recharge.	2-26;2-26A
	4. Defrost control out of adjustment.	4. Adjust defrost control.	2-62
	5. Defrost control sensing element not making proper contact.	5. Improve contact.	2-62
	6. Defective defrost control, time clock, or relay.	6. Replace control, time clock, or relay.	2-62
	7. Defective reversing valve.	7. Replace reversing valve.	2-61
	8. Wrong superheat setting on outdoor expansion valve.	8. Adjust superheat.	2-27
	9. Defective power element on outdoor expansion valve.	9. Replace power element.	2-27;2-27E
	10. Plugged outdoor expansion valve.	10. Clean or replace expansion valve.	2-27;2-27A, B,C,D,E
Ice buildup on lower section of outdoor coil.	1. Defective defrost relay.	1. Replace defrost relay.	2-65
	2. Defective compressor valves.	2. Replace valves and valve plate or compressor.	2-7;2-7C
	3. Shortage of refrigerant.	3. Repair leak and recharge.	2-26;2-26A
	4. Defrost control out of adjustment.	4. Adjust defrost control.	2-62
	5. Defrost sensing element not making proper contact.	5. Improve contact.	2-62
	6. Defective reversing valve.	6. Replace reversing valve.	2-61

(continued)

1-3 HEAT PUMP (HEATING CYCLE) (continued)

Condition	Possible cause	Corrective action	Reference
	7. Wrong superheat setting on outdoor expansion valve.	7. Adjust superheat.	2-27
Liquid refrigerant flooding compressor on heating cycle (TXV system).	1. Wrong superheat setting on outdoor expansion valve.	1. Adjust superheat.	2-27
	2. Outdoor expansion valve thermal bulb not making proper contact.	2. Improve contact.	2-27;2-27A
	3. Outdoor expansion valve stuck open.	3. Clean or replace expansion valve.	2-27;2-27B
	4. Leaking outdoor check valve.	4. Replace check valve.	2-60
Liquid refrigerant flooding compressor on heating cycle (capillary tube).	1. Refrigerant overcharge.	1. Purge overcharge.	2-25;2-25F
	2. High head pressure.	2. See "Head pressure" high (p. 18).	
	3. Defective outdoor check valve.	3. Replace check valve.	2-60
Excessive operating costs.	1. Refrigerant shortage.	1. Repair leak and recharge.	2-26;2-26A
	2. Defective reversing valve.	2. Replace reversing valve.	2-61
	3. Defrost control out of adjustment.	3. Adjust defrost control.	2-62
	4. Refrigerant overcharge.	4. Purge overcharge.	2-25;2-25F
	5. Dirty indoor or outdoor coil.	5. Clean coil.	2-26;2-26C
	6. Blower belt slipping on indoor or outdoor coil.	6. Replace or adjust belt.	2-25;2-25B
	7. Dirty indoor air filters.	7. Clean or replace filters.	2-26;2-26B
	8. Wrong thermostat location.	8. Relocate thermostat.	2-13;2-13C
	9. Ducts not insulated.	9. Insulate ducts.	2-40
	10. Wrong-size unit.	10. Replace with proper-size unit.	2-41
	11. Outdoor thermostat not controlling auxiliary heat.	11. Adjust, relocate, or provide shield.	2-13;2-13E

1-4 HEAT PUMP (HEATING OR COOLING CYCLE)

Condition	Possible cause	Corrective action	Reference
Compressor hums but will not start.	1. Faulty fuse.	1. Replace fuse and correct cause.	2-3
	2. Faulty wiring.	2. Repair wiring.	2-15
	3. Loose electrical terminals.	3. Repair loose connections.	2-15
	4. Compressor overload.	4. Locate and remove overload.	2-41
	5. Faulty starting capacitor.	5. Replace capacitor.	2-18;2-18A
	6. Faulty starting relay.	6. Replace relay.	2-19;2-19A, B,C,D,E,F
	7. Burned compressor motor.	7. Replace compressor.	2-35;2-35A, B,C,D,E,F
	8. Defective compressor bearings.	8. Replace bearings or compressor.	2-7;2-7B
	9. Stuck compressor.	9. Replace compressor.	2-7;2-7A,B
Compressor cycling on overload.	1. Low voltage.	1. Determine reason and repair.	2-17
	2. Loose electrical terminals.	2. Repair terminals.	2-15
	3. Single phasing of Three-phase power.	3. Replace fuse or repair wiring; or notify power company.	2-1;2-3;2-15
	4. Defective contactor contacts.	4. Replace contacts or contactor.	2-10;2-10C
	5. Defective compressor overload.	5. Replace overload.	2-6;2-6F,G,H, I,J,K,L
	6. Compressor overloaded.	6. Locate and remove overload.	2-5;2-6;2-7; 2-7A,B;2-8A,B; 2-41
	7. Defective start capacitor.	7. Replace capacitor.	2-18;2-18A
	8. Defective run capacitor.	8. Replace capacitor.	2-18;2-18B
	9. Defective starting relay.	9. Replace starting relay.	2-19;2-19A, B,C,D,E,F
	10. Refrigerant overcharge.	10. Purge refrigerant.	2-25;2-25F
	11. Defective compressor bearings.	11. Replace bearings or compressor.	2-7;2-7A,B
	12. Air or noncondensables in system (high head pressure).	12. Purge noncondensables from system.	2-25;2-25G
	13. Defective reversing valve.	13. Replace reversing valve.	2-61

(continued)

1-4 HEAT PUMP (HEATING OR COOLING CYCLE) (continued)

Condition	Possible cause	Corrective action	Reference
Compressor off on high-pressure control.	1. Refrigerant overcharge.	1. Purge overcharge.	2-25;2-25F
	2. Control out of adjustment.	2. Adjust control.	2-14;2-14B
	3. Defective indoor fan motor.	3. Repair or replace motor.	2-35;2-35A, B,C,D,E,F
	4. Defective outdoor fan motor.	4. Repair or replace motor.	2-35;2-35A, B,C,D,E,F
	5. Defective fan relay on either indoor or outdoor section.	5. Repair or replace relay.	2-39;2-42
	6. Defrost cycle too long.	6. Replace time clock, defrost relay, or termination thermostat.	2-62
	7. Defective reversing valve.	7. Replace reversing valve.	2-61
	8. Blower belt slipping on indoor or outdoor coil.	8. Adjust or replace belt.	2-26;2-26E
	9. Indoor or outdoor coil dirty.	9. Clean proper coil.	2-25;2-25B; 2-26;2-26C
	10. Dirty indoor air filters.	10. Replace or clean filters.	2-26;2-26B
	11. Air bypassing indoor or outdoor coil.	11. Prevent air bypassing.	2-63
	12. Air volume too low over indoor or outdoor coil.	12. Increase indoor ductwork or remove restriction from coils.	2-40
	13. Auxiliary heat strips ahead of indoor coil.	13. Locate heat strips downstream of indoor coil.	2-64
Compressor cycles on low-pressure control.	1. Refrigerant shortage.	1. Repair leak and recharge.	2-26;2-26A
	2. Low suction pressure.	2. Increase load (see "Suction pressure low," p. 18).	
	3. Defective expansion valve.	3. Repair or replace expansion valve.	2-27;2-27A, B,C,D,E
	4. Dirty indoor or outdoor coil.	4. Clean coil.	2-25;2-25B; 2-26;2-26C
	5. Slipping blower belt.	5. Replace or adjust belt.	2-25;2-25B; 2-26;2-26E
	6. Dirty indoor air filter.	6. Clean or replace filter.	2-26;2-26B
	7. Ductwork restriction.	7. Increase ductwork.	2-40

1–4 HEAT PUMP (HEATING OR COOLING CYCLE) (continued)

Condition	Possible cause	Corrective action	Reference
	8. Liquid drier or suction strainer restricted.	8. Replace drier or strainer.	2-28;2-28A, B,C,D,E
	9. Defrost thermostat element loose or making poor contact.	9. Tighten or increase contact.	2-60;2-61;2-62
	10. Air temperature too low for evaporation.	10. Relocate unit or provide adequate air temperature.	2-13;2-13A,B
	11. Defrost cycle too long.	11. Replace time clock, defrost relay, or termination thermostat.	2-62
	12. Defective evaporator fan motor.	12. Repair or replace fan motor.	2-35;2-35A, B,C,D,E,F
Outdoor fan runs but compressor will not.	1. Faulty electrical wiring or loose connections.	1. Repair wiring or connections.	2-15;2-16
	2. Defective starting capacitor.	2. Replace starting capacitor.	2-18;2-18A
	3. Defective starting relay.	3. Replace relay.	2-19;2-19A, B,C,D,E,F
	4. Defective run capacitor.	4. Replace run capacitor.	2-18;2-18B
	5. Shorted or grounded compressor motor.	5. Replace compressor.	2-6;2-6A, B,C,D,E
	6. Stuck compressor.	6. Replace compressor.	2-7;2-7A,B
	7. Compressor overloaded.	7. Determine and remove overload.	2-5;2-6;2-7; 2-7A,B;2-8; 2-8A,B;2-41
	8. Defective contactor contacts.	8. Replace contactor or contacts.	2-10;2-10B,C
	9. Single phasing of three-phase power.	9. Locate problem and repair or contact power company.	2-3;2-15
	10. Low voltage.	10. Locate and correct cause.	2-1
Outdoor fan motor will not start.	1. Faulty electrical wiring or loose connections.	1. Repair wiring or connections.	2-15;2-16
	2. Defective outdoor fan motor.	2. Repair or replace motor.	2-35;2-35A, B,C,D,E,F
	3. Defective outdoor fan relay.	3. Replace fan relay.	2-42

(continued)

1-4 HEAT PUMP (HEATING OR COOLING CYCLE) (continued)

Condition	Possible cause	Corrective action	Reference
	4. Defective defrost control, timer, or relay.	4. Replace control, timer, or relay.	2-62
Outdoor section does not run.	1. No electrical power.	1. Inform power company.	2-1
	2. Blown fuse.	2. Replace fuse and correct fault.	2-3
	3. Faulty electrical wiring or loose terminals.	3. Repair wiring or terminals.	2-15;2-16
	4. Compressor overloaded.	4. Determine overload and correct.	2-5;2-6; 2-7A,B;2-8A, B;2-41
	5. Defective transformer.	5. Replace transformer.	2-36;2-36A
	6. Burned contactor coil.	6. Replace contactor coil.	2-10;2-10A
	7. Compressor overload open.	7. Determine cause and correct.	2-6;2-6F,G, H,I,J,K,L
	8. High-pressure control open.	8. Determine cause and correct.	2-14;2-14B
	9. Low-pressure control open.	9. Determine cause and correct.	2-14;2-14A
	10. Thermostat off.	10. Turn thermostat on and set.	2-13;2-13D
Indoor blower will not run.	1. Blown fuse.	1. Replace fuse and correct cause.	2-3
	2. Faulty electrical wiring or loose connections.	2. Repair wiring or connections.	2-15;2-16
	3. Burned transformer.	3. Replace transformer.	2-36;2-36A
	4. Indoor fan relay defective.	4. Replace fan relay.	2-39
	5. Faulty indoor fan motor.	5. Repair or replace motor.	2-35;2-35A, B,C,D,E,F
	6. Faulty thermostat.	6. Replace thermostat.	2-13;2-13A, B,C,D,E
Indoor coil iced over.	1. Dirty filters.	1. Clean or replace filters.	2-25;2-25B; 2-26;2-26B
	2. Dirty coil.	2. Clean coil.	2-26;2-26C
	3. Blower fan belt slipping.	3. Replace or adjust belt.	2-26;2-26E
	4. Outdoor check valve sticking closed.	4. Replace check valve.	2-60
	5. Defective indoor expansion valve.	5. Clean or replace expansion valve.	2-27;2-27A, B,C,D,E
	6. Low indoor air temperature.	6. Increase temperature.	2-26

1-4 HEAT PUMP (HEATING OR COOLING CYCLE) (continued)

Condition	Possible cause	Corrective action	Reference
	7. Shortage of refrigerant.	7. Repair leak and recharge.	2-26;2-26A
Noisy compressor.	1. Low oil level in compressor.	1. Determine reason for loss of oil and correct. Replace oil.	2-26;2-26A, B,C,D,E,F; 2-27;2-27B; 2-28;2-28A, B,C,D,E;2-31; 2-31A,B,C; 2-32;2-32A
	2. Defective suction and/or discharge valves.	2. Replace valves and valve plate or compressor.	2-7;2-7C
	3. Loose hold-down bolts.	3. Tighten.	2-30
	4. Broken internal springs.	4. Replace compressor.	2-5;2-8;2-30
	5. Inoperative check valves.	5. Repair or replace check valves.	2-60
	6. Loose thermal bulb on indoor expansion valve.	6. Tighten thermal bulb.	2-27;2-27A
	7. Improper superheat setting on indoor expansion valve.	7. Adjust superheat.	2-27
	8. Stuck-open indoor expansion valve.	8. Clean or replace expansion valve.	2-27;2-27B
Compressor loses oil.	1. Refrigerant shortage.	1. Repair leak and recharge.	2-26;2-26A
	2. Low suction pressure.	2. Increase load on evaporator.	2-26;2-26A, B,C,D,E,F
	3. Restriction in refrigerant circuit.	3. Remove restriction.	2-28;2-28A, B,C,D,E
Unit operates normally in one cycle, but high suction pressure on other cycle.	1. Leaking check valve.	1. Replace check valve.	2-60
	2. Loose thermal bulb on outdoor or indoor expansion valve.	2. Tighten thermal bulb.	2-27;2-27A
	3. Leaking reversing valve.	3. Replace reversing valve.	2-61
	4. Expansion valve stuck open on indoor or outdoor coil.	4. Repair or replace expansion valve.	2-27;2-27B
Unit pumps down in cool or defrost cycle but	1. Defective reversing valve.	1. Replace reversing valve.	2-61
	2. Defective power element on indoor expansion valve.	2. Replace power element.	2-27;2-27E

(continued)

1–4 HEAT PUMP (HEATING OR COOLING CYCLE) (continued)

Condition	Possible cause	Corrective action	Reference
operates normally in heat cycle.	3. Restriction in refrigerant circuit.	3. Locate and remove restriction.	2-28;2-28A, B,C,D,E
	4. Clogged indoor expansion valve.	4. Clean or replace expansion valve.	2-27;2-27E
	5. Check valve in outdoor section sticking closed.	5. Replace check valve.	2-60
Head pressure high.	1. Overcharge of refrigerant.	1. Purge overcharge.	2-25;2-25F
	2. Air or noncondensables in system.	2. Purge noncondensables.	2-25;2-25G
	3. High air temperature.	3. Reduce air temperature.	2-37
	4. Dirty indoor or outdoor coil.	4. Clean coil.	2-25;2-25B; 2-26;2-26C
	5. Dirty indoor air filters.	5. Clean or replace filters.	2-26;2-26B,C
	6. Indoor or outdoor blower belt slipping.	6. Replace or adjust blower belt.	2-25;2-25B
	7. Air bypassing indoor or outdoor coil.	7. Prevent air bypassing.	2-63
Suction pressure high.	1. Defective compressor valves.	1. Replace valves and valve plate or compressor.	2-7;2-7C
	2. High head pressure.	2. See "Head pressure high" (above).	
	3. Excessive load on cooling.	3. Determine cause and correct.	2-41
	4. Leaking reversing valve.	4. Replace reversing valve.	2-61
	5. Leaking check valves.	5. Replace check valves.	2-60
	6. Indoor or outdoor expansion valve stuck open.	6. Clean or replace expansion valve.	2-27;2-27B
	7. Loose thermal bulb on indoor or outdoor expansion valve.	7. Tighten bulb.	2-27;2-27A
Suction pressure low.	1. Shortage of refrigerant.	1. Repair leak and recharge.	2-26;2-26A
	2. Blower belt slipping on indoor or outdoor blower.	2. Replace or adjust belt.	2-26;2-26E
	3. Dirty indoor air filters.	3. Clean or replace filters.	2-26;2-26B
	4. Defective check valves.	4. Replace check valves.	2-60

1-4 HEAT PUMP (HEATING OR COOLING CYCLE) (continued)

Condition	Possible cause	Corrective action	Reference
	5. Restriction in refrigerant circuit.	5. Locate and remove restriction.	2-28;2-28A, B,C,D,E
	6. Ductwork small or restricted.	6. Repair or replace ductwork.	2-40
	7. Defective expansion valve power element on indoor or outdoor coil.	7. Replace power element.	2-27;2-27E
	8. Clogged indoor or outdoor expansion valve.	8. Clean or replace valve.	2-27;2-27E
	9. Wrong superheat setting on indoor or outdoor expansion valve.	9. Adjust superheat setting.	2-27
	10. Dirty indoor or outdoor coil.	10. Clean coil.	2-25;2-25B; 2-26;2-26C
	11. Bad contactor contacts.	11. Replace contactor or contacts.	2-10;2-10C
	12. Low refrigerant charge.	12. Repair leak and recharge.	2-26;2-26A

1-5 OIL HEAT

Condition	Possible cause	Corrective action	Reference
Burner will not start.	1. Thermostat off or set too low.	1. Turn thermostat on or set to higher temperature.	2-13;2-13D
	2. Burner motor overload tripped.	2. Push motor overload reset button.	2-6H,L
	3. Primary control off on safety.	3. Reset safety switch.	2-71
	4. Dirty thermostat contacts.	4. Clean thermostat contacts.	2-13;2-13A
	5. Bad thermostat.	5. Replace thermostat.	2-13;2-13A,B
	6. Blown fuse or tripped breaker.	6. Replace fuse or reset breaker.	2-3
	7. Disconnect switch open.	7. Close switch.	2-2
	8. Shorted flame detector circuit.	8. Replace flame detector.	2-66
	9. Shorted flame detector leads.	9. Separate and insulate leads.	2-67

(continued)

1-5   OIL HEAT (continued)

Condition	Possible cause	Corrective action	Reference
	10. Flame detector exposed to direct light.	10. Protect detector from light.	2-66
	11. Faulty friction clutch.	11. Replace element or control.	2-68
	12. Hot contacts stuck.	12. Replace element or control.	2-69
	13. Dirty cold contacts.	13. Clean contacts.	2-69
	14. Flame detector bimetal carboned.	14. Clean bimetal.	2-70
	15. Loose connection or broken wire on flame detector.	15. Repair connection or replace wire.	2-15
	16. Low line voltage or power failure.	16. Notify power company.	2-1
	17. Limit control open.	17. Set limit to 200°F (93°C), then jumper control terminals; if burner starts, replace the control.	2-58
	18. Open electric circuit to limit control.	18. Repair or replace wiring.	2-15
	19. Defective internal primary control circuit.	19. Replace control.	2-71
	20. Dirty burner relay contacts in primary control.	20. Clean contacts.	2-71
	21. Defective burner motor.	21. Replace burner motor.	2-35;2-35A, B,C,D,E,F
	22. Binding burner blower wheel.	22. Turn off power and rotate blower by hand; if binding, free.	2-35;2-35A, B,C,D,E,F 2-72
	23. Seized fuel pump.	23. Turn power off and rotate blower by hand; if binding, replace fuel pump.	2-73
Burner starts and fires but short cycles.	1. Thermostat in warm draft.	1. Relocate thermostat.	2-13;2-13C
	2. Heat anticipator set wrong.	2. Correct anticipator setting.	2-13;2-13B
	3. Furnace blower running too slow.	3. Speed up blower to obtain a temperature rise of 85 to 95°F (29.4 to 35°C).	2-47;2-48; 2-48A
	4. Limit control set too low.	4. Reset limit to 200°F (93°C).	2-58

1-5   OIL HEAT (continued)

Condition	Possible cause	Corrective action	Reference
	5. Dirty air filter.	5. Clean or replace filter.	2-26B
	6. Return air restriction.	6. Clear restriction.	
	7. Low or fluctuating voltage.	7. Notify power company.	2-1
	8. Loose wiring connection.	8. Repair connection.	2-15
Burner starts and fires but does not heat enough (short cycles).	1. Vibration at thermostat.	1. Correct vibration or relocate thermostat.	2-13;2-13C
	2. Thermostat in warm draft.	2. Relocate thermostat.	2-13;2-13C
	3. Heat anticipator set wrong.	3. Correct anticipator setting.	2-13;2-13B
	4. Furnace blower running too slow.	4. Speed up blower to obtain a temperature rise to 85 to 95°F (29.4 to 35°C).	2-47;2-48A
	5. Dirty air filter.	5. Clean or replace filter.	2-26B
	6. Defective blower motor bearings.	6. Replace motor.	2-35;2-35E,F
	7. Defective blower bearings.	7. Replace bearings.	2-48;2-48B
	8. Dirty furnace blower wheel.	8. Clean blower wheel.	2-26D
	9. Wrong blower motor rotation.	9. Change rotation or replace motor.	2-48
	10. Return air restricted.	10. Clear restriction.	2-40
	11. Limit control set too low.	11. Reset limit to 200°F (93°C).	2-58
	12. Low or fluctuating voltage.	12. Notify power company.	2-1
	13. Loose wiring connection.	13. Repair connection.	2-15
Burner starts and fires; then locks out on safety.	1. Too little primary air; long dirty flame.	1. Increase combustion air.	2-74
	2. Too much primary air; short lean flame.	2. Reduce combustion air.	2-74
	3. Unbalanced flame.	3. Replace nozzle.	2-77
	4. Too little or restricted draft.	4. Correct draft or remove restriction.	2-74
	5. Excessive draft.	5. Adjust barometric damper.	2-74
	6. Dirty flame detector bimetal element.	6. Clean element.	2-70

(continued)

1-5 OIL HEAT (continued)

Condition	Possible cause	Corrective action	Reference
	7. Faulty flame detector clutch.	7. Replace flame detector control.	2-68
	8. Welded or shorted cold contacts in flame detector.	8. Replace flame detector control.	2-69
	9. Air leaking into flue pipe around flame detector mount.	9. Seal air leaks.	2-71
	10. Dirty flame detector cad cell face.	10. Clean cad cell face.	2-66
	11. Loose or defective flame detector cad cell wires.	11. Repair or replace cad cell holder and wires.	2-66
	12. Faulty flame detector cad cell; resistance exceeds 1500 Ω.	12. Replace cad cell.	2-66
	13. Defective primary control circuit.	13. Replace primary control.	2-71
Burner starts, but no flame is established.	1. Oil tank empty.	1. Contact oil distributor.	
	2. Oil tank shutoff valve closed.	2. Open valve.	
	3. Water in oil tank.	3. Remove water.	2-75
	4. Air leak in oil supply line.	4. Repair leak.	2-76
	5. Oil filter plugged.	5. Install new filter.	2-78
	6. Oil pump strainer plugged.	6. Clean strainer.	2-73
	7. Restricted oil line.	7. Repair or replace line.	2-76
	8. Excessive combustion air.	8. Adjust air supply damper.	2-74
	9. Excessive vent draft.	9. Adjust barometric damper to 0.030 in. water column.	2-74
	10. Off-center spray from nozzle.	10. Replace nozzle.	2-77
	11. Nozzle strainer plugged.	11. Replace nozzle.	2-78
	12. Nozzle orifice plugged.	12. Replace nozzle.	2-77
	13. Faulty oil pump.	13. Replace pump.	2-73
	14. Low fuel pressure.	14. Adjust pressure to desired pressure.	2-73
	15. Faulty pump coupling.	15. Replace coupling.	2-73
	16. Low line voltage to transformer primary.	16. Notify power company.	2-1

1-5 OIL HEAT (continued)

Condition	Possible cause	Corrective action	Reference
	17. Faulty transformer.	17. Replace transformer.	2-79
	18. No or weak ignition spark.	18. Properly ground transformer case.	2-79
	19. Dirty or shorted ignition electrodes.	19. Clean electrodes.	2-79
	20. Improper position or gap of ignition electrodes.	20. Correctly position and reset electrodes.	2-79
	21. Cracked or burned lead insulation.	21. Replace electrode leads.	2-79
	22. Loose or disconnected electrode leads.	22. Repair or replace leads.	2-79
	23. Defective electrode lead insulators.	23. Replace electrodes.	2-79
	24. Oil pump or blower overloading motor.	24. Remove overload condition.	2-72;2-73
	25. Faulty oil pump motor.	25. Replace motor.	2-6;2-6A,B,C
	26. Low voltage.	26. Notify power company.	2-1
Burner starts and fires but loses flame and locks out on safety.	1. Dirty face and cad cell.	1. Clean cad cell face.	2-66
	2. Faulty cad cell; resistance exceeds 1500 Ω.	2. Replace cad cell.	2-66
	3. Loose or defective cad cell wires.	3. Repair or replace wires.	2-67
	4. Stack control bimetal dirty.	4. Clean bimetal element.	2-70
	5. Faulty friction clutch in stack control.	5. Replace stack control.	2-68
	6. Air leaking into vent around stack control mount.	6. Seal air leaks.	2-71
	7. Defective stack control cold contacts.	7. Replace stack control.	2-69
	8. Too much combustion air.	8. Adjust combustion air damper.	2-74
	9. Too little combustion air.	9. Adjust combustion air damper.	2-74
	10. Unbalanced flame.	10. Replace nozzle.	2-77
	11. Excessive vent.	11. Adjust barometric damper.	2-74
	12. Too little vent draft.	12. Adjust barometric damper.	2-74

(continued)

1-5 OIL HEAT (continued)

Condition	Possible cause	Corrective action	Reference
	13. Vent restricted.	13. Clear restriction.	2-74
	14. Oil pump loses prime.	14. Prime pump at bleed port.	2-73
	15. Air leak in oil supply line.	15. Repair leaks.	2-76
	16. Partially plugged nozzle.	16. Replace nozzle.	2-77
	17. Partially plugged nozzle strainer.	17. Replace nozzle.	2-77
	18. Water in oil storage tank.	18. Remove water from storage tank.	2-75
	19. Oil too heavy.	19. Change to No. 1 oil.	2-80
	20. Plugged fuel pump strainer.	20. Clean strainer or replace pump.	2-73
	21. Restricted oil line.	21. Clear restriction.	2-76
Too much heat; burner runs continuously.	1. Defective thermostat.	1. Repair or replace thermostat.	2-13
	2. Shorted thermostat wires.	2. Repair or replace wires.	2-13;2-15
	3. Thermostat in cold location.	3. Relocate thermostat.	2-13;2-13C
	4. Thermostat not level.	4. Level thermostat.	2-13
	5. Defective primary control.	5. Replace control.	2-71
Too little heat; burner runs continuously.	1. Too much combustion air.	1. Reduce combustion air.	2-74
	2. Air leaking into heat exchanger.	2. Repair leaks.	2-74
	3. Excessive vent draft.	3. Adjust barometric damper.	2-74
	4. Wrong burner head adjustment.	4. Correct burner head setting.	2-74
	5. Plugged heat exchanger.	5. Clean heat exchanger.	2-74
	6. Too little combustion air.	6. Increase combustion air.	2-74
	7. Insufficient vent draft.	7. Adjust barometric damper.	2-74
	8. Insufficient indoor air.	8. Speed blower to obtain a temperature rise to 85 to 95°F (29.4 to 35°C).	2-47;2-48; 2-48A
	9. Dirty indoor blower.	9. Clean blower.	2-26D;2-47 2-48
	10. Dirty furnace filter.	10. Clean or replace filter.	2-26B
	11. Partially plugged nozzle.	11. Replace nozzle.	2-77

1-5 *OIL HEAT (continued)*

Condition	Possible cause	Corrective action	Reference
	12. Nozzle too small.	12. Replace with larger nozzle.	2-77
	13. Low oil pressure.	13. Increase to proper pressure.	2-73
No oil flow at nozzle.	1. Oil level below intake line in supply tank.	1. Fill tank with oil.	2-75
	2. Clogged filter or strainer.	2. Remove and clean strainer. Replace filter element.	2-76; 2-78
	3. Clogged nozzle.	3. Replace nozzle.	2-77
	4. Air leak in intake line.	4. Repair leak.	
	5. Restricted intake line (high vacuum).	5. Remove restriction.	2-76
	6. A two-pipe system becomes air bound.	6. Remove air lock.	2-76
	7. A single-pipe system that becomes air bound.	7. Remove air lock.	2-76
	8. Slipping or broken coupling.	8. Tighten or replace coupling.	2-73
	9. Frozen pump shaft.	9. Replace pump.	2-73
Oil leak.	1. Loose plugs or fittings.	1. Dope with good-quality thread sealer, retighten.	2-76
	2. Leak at pressure adjustment screw or nozzle plug.	2. Washer may be damaged.	2-73
	3. Blown seal (single-pipe system.)	3. Replace oil pump.	2-73
	4. Blown seal (two-pipe system).	4. Replace oil pump.	2-73
	5. Seal leaking.	5. Replace oil pump.	2-73
	6. Cover leaking.	6. Tighten cover or replace damaged gasket.	2-72
Noisy operation.	1. Bad coupling	1. Align coupling.	
	2. Air in inlet line.	2. Repair leak.	2-76
	3. Tank hum on two-pipe system and inside tank.	3. Unstall return line hum eliminator in return line.	2-75; 2-76
Pulsating pressure.	1. Partially clogged strainer or filter.	1. Remove and clean strainer. Replace filter element.	2-78
	2. Air leak in intake line.	2. Repair leak.	2-76

(continued)

1-5   OIL HEAT (continued)

Condition	Possible cause	Corrective action	Reference
	3. Air leaking around cover.	3. Be sure that strainer cover screws are tightened securely. Check for damaged cover gasket.	2-72
Improper nozzle cutoff.	1. Filter leaks.	1. Check face of cover and gasket for damage.	2-78
	2. Strainer cover loose.	2. Tighten four screws on cover.	2-76
	3. Air pocket between cutoff valve and nozzle.	3. Remove air pocket.	2-76
	4. Air leak in intake line.	4. Repair leak.	2-76
	5. Partially clogged nozzle strainer.	5. Clean strainer or change nozzle.	2-78
	6. Leak at nozzle adapter.	6. Change nozzle and adapter.	2-77

2 Component Troubleshooting

This book is intended primarily to assist the service technician in diagnosing problems and repairing heating equipment. Use of this handbook should allow more competent and economical service of the equipment.

The problems encountered in servicing heating equipment can be divided into the following categories:

1. When the unit does not operate at all, the problem is in the electrical circuit.
2. When the system will run but will not heat, the problem is in the mechanical components.
3. The problem can also be a combination of electrical and mechanical malfunctions.

Therefore, if the service technician can determine whether the trouble is electrical or mechanical in nature, he or she has eliminated half of the possible causes of the problem. An example of a combination of electrical and mechanical problems would be if the bearings in a compressor became stuck and caused the compressor motor to burn out.

2-1 POWER FAILURE

When a power failure occurs, the electric company must be contacted to make the necessary repairs. The repair may range from replacing a fuse on an electric pole to replacing a transformer, but no matter what it is, it is the responsibility of the electric company. After the repair has been made, the heating service technician should check all equipment to be certain that it is operating properly. Sometimes an electrical failure will cause damage to electric motors and they will need to be repaired or replaced.

2-2 DISCONNECT SWITCHES

Disconnect switches are electrical switches which are used to interrupt electrical power to electric furnaces, outdoor units on heat pump systems, large electric motors, and other equipment that requires heavy current flow. These switches are usually installed within an arm's reach of the equipment that they control. Some of these switches contain only a set of electrical contacts, whereas others contain fuses to the equipment in addition to the contacts. Located on the outside of the switch box is a lever used to open and close the contacts (see Fig. 2-1). This lever may accidentally be bumped, causing the contacts to open, or it may accidentally be left in the "off" position. In either case the lever must be returned to the "on" position before operation can be resumed.

2-3 FUSES

An electrical fuse is a protective device placed in the electrical line to an electric circuit. There are basically two types of fuses: the plug type and the cartridge type (see Fig. 2-2). The purpose of a fuse is to protect the electric circuit in case of an electrical overload. This overload may be due to tight bearings, shorted motor winding, breakdown of electrical insulation, tight fan belt, burned contactor contacts,

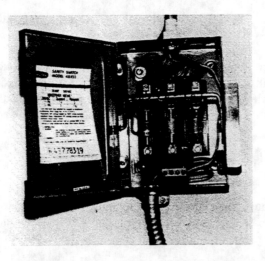

Figure 2-1. Mounted disconnect box.

Buss one-time fuses
Non-renewable

1 to 60 ampere   100 ampere

Buss Fustat fuses and
adapters

Dual element - Time delay - Type S base

For motor protection, take amperage of motor and
add 25%. Adapter makes Fustat fuses fit any plug
fuse holder.

Buss clear window plug fuses

Figure 2-2   Plug- and cartridge-type fuses. ("Reprinted with permission by Buss-mann Division, Cooper Industries.")

and so on. The problem that caused the blown fuse must be found and eliminated before the job is complete.

Defective fuses may be found with either a voltmeter or an ohmmeter. To check fuses with a voltmeter, select a scale that is high enough to prevent damage to the meter. Check the voltage across each fuse (see Fig. 2-3). If a voltage is found between the ends of the fuse, it is defective and must be replaced with one of the proper size. Obviously, this procedure requires that the electrical power be on. Use caution to avoid electrical shock.

To check with an ohmmeter, remove the fuse from the holders and check for continuity between the ends of the fuse (see Fig. 2-4). If no continuity is found, replace the fuse with one of the proper size.

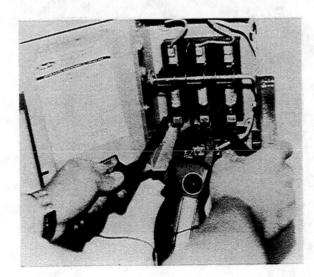

Figure 2-3. Checking fuses in a disconnect box.

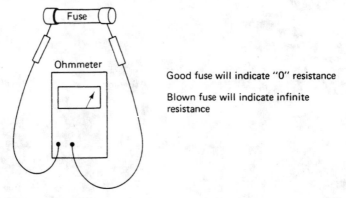

Good fuse will indicate "0" resistance

Blown fuse will indicate infinite resistance

Figure 2-4. Checking fuses out of disconnect box with ohmmeter.

2-4 CIRCUIT BREAKERS

Circuit breakers are protective devices that are used to protect the electrical circuit in much the same manner as the fuse. They may be used for disconnecting the circuit as well as for circuit protection (see Fig. 2-5). Circuit breakers are available in ratings for use with small household refrigerators as well as large commercial installations.

To check a circuit breaker, reset the breaker to the "on" position and check the voltage between the line connection to the breaker and the ground terminal inside the breaker box (see Fig. 2-6). Some breakers require that the reset lever be pushed farther toward the "off" position to reset them. If voltage is indicated at this point, the breaker is reset and is probably good. However, if the breaker con-

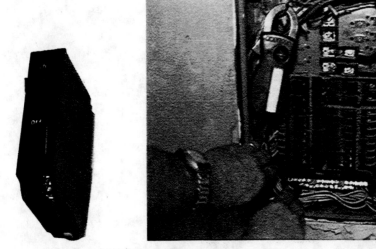

Figure 2-5. Circuit breaker. **Figure 2-6.** Checking circuit breaker inside a box.

tinues to trip, check the current draw through the circuit with an ammeter. If the current draw is within the rating of the circuit breaker and it continues to trip, the breaker is weak and should be replaced.

2-5 COMPRESSOR

The compressor is a device used to circulate the refrigerant through the system (see Fig. 2-7). It has two functions: (1) It draws the refrigerant vapor from the evaporator and lowers the pressure of the refrigerant in the evaporator to the desired evaporating temperature; and (2) the compressor raises the pressure of the refrigerant vapor in the condenser high enough so that the saturation temperature is higher than the temperature of the cooling medium used to cool the condenser and condense the refrigerant.

The compressors in use today are usually of the hermetic or semihermetic type. Therefore, the problems encountered could be electrical, mechanical, or a combination of the two.

2-6 COMPRESSOR ELECTRICAL PROBLEMS

The problems encountered in the electrical circuit of a compressor may be divided into the following classifications: open winding, shorted winding, or grounded winding. An accurate ohmmeter is needed to check for these conditions. The following checks are good for any type of electric motor.

(a)   (b)

Figure 2-7.   (a) Semihermetic compressor; (b) hermetic compressor.

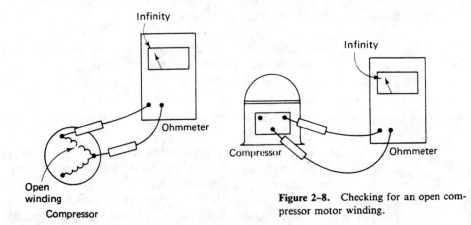

Figure 2–8. Checking for an open compressor motor winding.

A. Open Compressor Motor Windings

Open compressor motor windings occur when the path for electrical current is interrupted. This interruption occurs when the wire insulation becomes bad and allows the wire to overheat and burn apart. To check for an open winding, remove all external wiring from the motor terminals. Using an ohmmeter, check the continuity from one terminal to another terminal (see Fig. 2–8). Be sure to zero the ohmmeter. There should be no continuity from any terminal to the motor case.

B. Shorted Compressor Motor Windings

Shorted compressor motor windings occur when the insulation on the winding becomes bad and allows a shorted condition (two wires touch), which allows the electrical current to bypass part of the winding (see Fig. 2–9). In some instances,

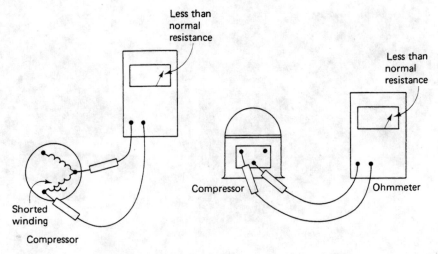

Figure 2–9. Checking for a shorted compressor winding.

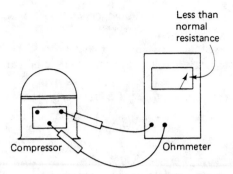

Figure 2-10. Checking compressor motor winding resistance.

depending on how much of the winding is bypassed, the motor may continue to operate but will draw excessive amperage. To check for a shorted winding, remove all external wiring from the motor terminals. Using an ohmmeter, check the continuity from one terminal to another terminal. Be sure to zero the ohmmeter (see Fig. 2-10). The shorted winding will be indicated by a less-than-normal resistance. In some cases it will be necessary to consult the manufacturer's data for that particular motor to determine the correct resistance requirements. There should be no continuity from any terminal to the motor case.

C. Grounded Compressor Motor Windings

Grounded compressor motor windings occur when the insulation on the winding is broken down and the winding becomes shorted to the housing (see Fig. 2-11). In such cases the motor will rarely run and will immediately blow fuses or trip the circuit breaker. To check for a grounded winding, remove all external wiring from the motor terminals. Using the ohmmeter, check the continuity from each terminal to the motor case. Be sure to zero the ohmmeter (see Fig. 2-12). A grounded winding

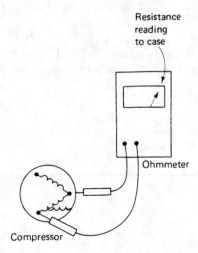

Figure 2-11. Grounded compressor motor winding.

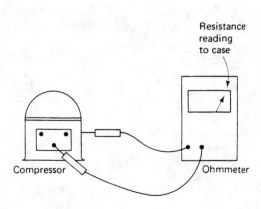

Figure 2-12. Checking for a grounded compressor motor winding.

will be indicated by a low-resistance reading. It may be necessary to remove some paint or scale from the motor case so that an accurate reading can be obtained.

D. Determining the Common, Run, and Start Terminals

Determining the common, run, and start terminals of a compressor is a relatively simple process. Be sure that all external wiring is removed from the compressor terminals so that no false readings are indicated. Draw the terminal configuration on a piece of paper. Measure the resistance between each terminal with an ohmmeter. Be sure to zero the ohmmeter (see Fig. 2–13). Apply the following formula: The least resistance indicated is between the run and common terminals; the medium resistance indicated is between the common and start terminals; and the most resistance indicated is between the start and run terminals. The compressor motor should be properly wired.

E. Checking Two-Speed Compressor Windings

Checking two-speed compressor windings is a relatively simple process. The first step is to turn off all electrical power to the unit, then remove all external wiring from the compressor motor terminals. Set the ohmmeter on the R × 10K scale and zero the ohmmeter. Then check the windings for grounds by touching one probe of the ohmmeter to each terminal and the other probe to the compressor housing. Any resistance reading indicates that there is a grounded winding.

To check for *open windings* in a two-speed single-phase compressor motor, refer to Fig. 2–14. Check for continuity between terminals 7 and 3, between terminals 7 and 8, and between terminals 7 and 2. When no continuity is indicated with either of these checks, there is an open winding and the compressor must be replaced.

To check for *open windings* in a two-speed three-phase compressor motor, refer to Fig. 2–15. Check for continuity between terminals 1 and 3, and between terminals 1 and 2. When no continuity is indicated with either of these checks, there is an open winding and the compressor must be replaced.

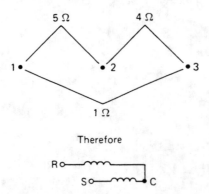

Therefore

Figure 2–13. Locating compressor terminals.

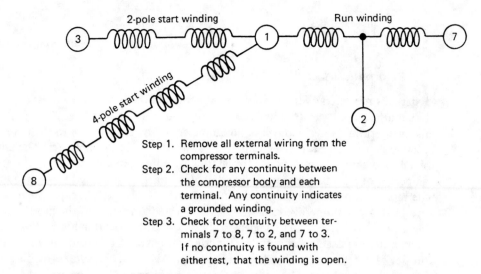

Step 1. Remove all external wiring from the compressor terminals.

Step 2. Check for any continuity between the compressor body and each terminal. Any continuity indicates a grounded winding.

Step 3. Check for continuity between terminals 7 to 8, 7 to 2, and 7 to 3. If no continuity is found with either test, that the winding is open.

Figure 2–14. Checking the motor windings in a two-speed, single-phase compressor motor.

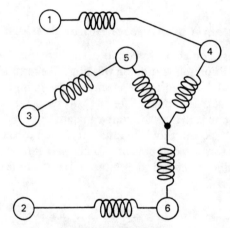

Step 1. Remove all external wiring from the compressor terminals.

Step 2. Check for any continuity between the compressor body and each terminal. Any continuity indicates a grounded winding.

Step 3. Check for continuity between terminals 1 to 2 and 1 to 3. If no continuity is found with either test, that the winding is open.

Figure 2–15. Checking the motor windings in a two-speed, three-phase compressor motor.

F. Compressor Motor Overloads

Compressor motor overloads are used to protect the compressor motor from damage that might occur from overcurrent, overtemperature, or both. Motor overloads may be mounted externally or internally, depending on the design of the compressor. They are generally mounted near the hottest part of the motor winding.

G. Externally Mounted Overloads

Externally mounted overloads are manufactured in three configurations: (1) the two-terminal, (2) the three-terminal, and (3) the four-terminal (see Fig. 2–16). To check the two-terminal overload, place an ammeter on the electric line to the compressor common terminal. Start the compressor while observing the ammeter. The ammeter should indicate a momentary current flow of approximately six times the running amperage of the compressor motor, then drop back to the rated amperage draw of the motor or below. If the overload then cycles the motor, the problem is in the overload. If the amperage remains above the rated amperage of the motor, the trouble is not in the overload. To check to be sure that the overload is cycling the compressor motor, check across the overload terminals with a voltmeter while the compressor is off (see Fig. 2–17). If the overload is open, a voltage reading will be indicated. No voltage reading indicates that the overload has not opened the circuit. The trouble is elsewhere. Be sure to replace these overloads with the type recommended by the manufacturer.

H. Three-Terminal External Compressor Motor Overload

This type of overload is used on compressor motors, where it is desirable to protect the starting winding in addition to the running winding. The terminals are numbered 1, 2, and 3. Terminal 1 is connected to the electrical line going to the compressor common terminal, terminal 2 is connected to the run terminal of the compressor motor, and terminal 3 is connected to the start capacitor.

    Connections of this type will provide closer protection of the compressor motor in the event of a bad-starting component. To check a three-terminal overload, place an ammeter on the line connected to terminal 1 and start the compressor while

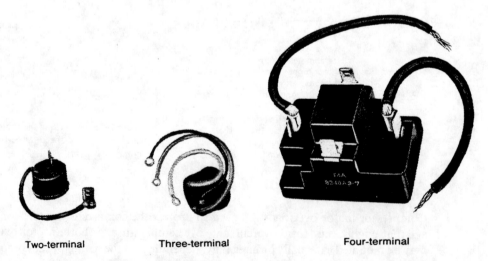

Two-terminal   Three-terminal   Four-terminal

Figure 2–16.   Two-, three-, and four-terminal compressor overload protectors.

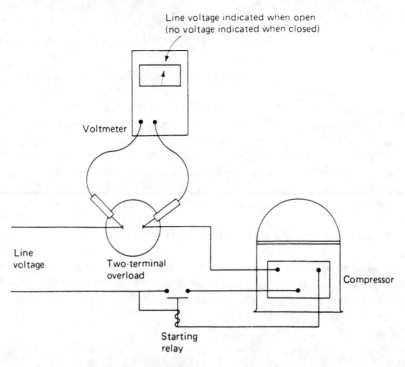

Line voltage indicated when open
(no voltage indicated when closed)

Voltmeter

Line
voltage

Two-terminal
overload

Starting
relay

Compressor

Figure 2–17. Checking voltage across a two-terminal overload.

observing the ammeter (see Fig. 2–18). The ammeter should indicate a momentary current flow of approximately six times the normal running current of the motor, then drop back to the amperage rating of the motor or below. If the overload then cycles the compressor motor, the problem is in the overload. If the amperage draw

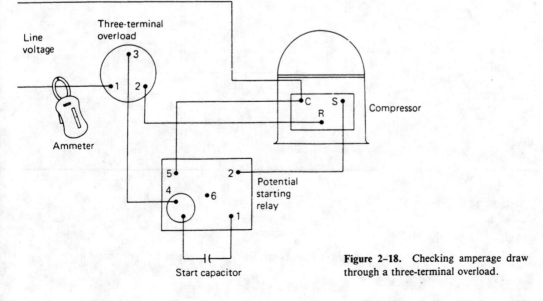

Line
voltage

Three-terminal
overload

Ammeter

Compressor

Potential
starting
relay

Start capacitor

Figure 2–18. Checking amperage draw through a three-terminal overload.

remains above the rated amperage of the motor, the trouble is not in the overload. To check to see if the trouble is in the starting or running circuit, measure the amperage draw through the wire connected to terminal 2, then the wire to terminal 3, while the compressor is running. The circuit with the high amperage draw is where the fault is. The external components must be checked. To be certain that the overload is cycling the compressor motor, check the voltage between terminals 1 and 2, then between terminals 1 and 3, while the compressor is off and cool. If voltage is indicated, replace the overload (see Fig. 2–19). If the overload must be replaced, be sure to use an exact replacement to provide the proper protection.

I. Four-Terminal External Compressor Motor Overload

This type of motor overload is used on larger compressor motors and is usually mounted away from the compressor in the control panel or on the starter. This type of overload senses only the current draw of the compressor motor (see Fig. 2–20).

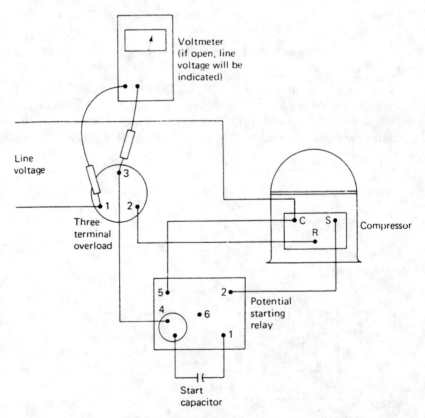

Figure 2–19.   Checking voltage across terminals 1 and 3 on a three-terminal overload.

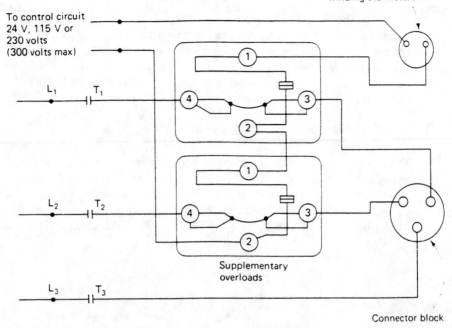

Figure 2-20. External thermostat overload connection.

The four-terminal overload is actuated by a bimetal, by the melting of a special type of solder, or by a hydraulic fluid in a cylinder. The bimetal and solder types will have two electrical connections for the compressor motor and two electrical connections for the control circuit (see Fig. 2-21).

When the current flow to the compressor motor remains over the rating of the overload for a definite period of time, the bimetal or solder pot will become heated and will allow the control circuit to open. This in turn interrupts the control circuit, which in turn stops the compressor motor. To check these overloads, place an ammeter on the compressor motor common wire and start the compressor while observing the ammeter. The ammeter should indicate a momentary flow of approximately six times the running amperage of the compressor motor, then drop back to the rated amperage of the motor or below. If the overload then cycles, the problem is in the overload. If the amperage remains above the rated amperage of the motor, the trouble is not in the overload. To check to make sure the overload

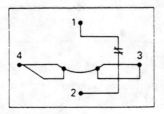

Figure 2-21. Internal view of a four-terminal overload.

is cycling the compressor motor, check across the control circuit terminals of the overload with a voltmeter while the compressor is off and cool (see Fig. 2–22). If the overload is open, a voltage will be indicated. No voltage reading indicates that the overload has not opened the circuit. The trouble is elsewhere. These overloads may be either manual or automatic reset types. Be sure to replace them with an exact replacement so that proper motor protection will be maintained.

J. Hydraulic-Fluid Compressor Motor Overload

This type of overload operates on the current flow to the compressor motor only and is generally used on larger compressor motors. It is located in the control panel away from the compressor motor. The current passing through the overload is routed through a heater, which causes the hydraulic fluid to become warm. This warming action increases the pressure of the fluid, causing the electric circuit to the com-

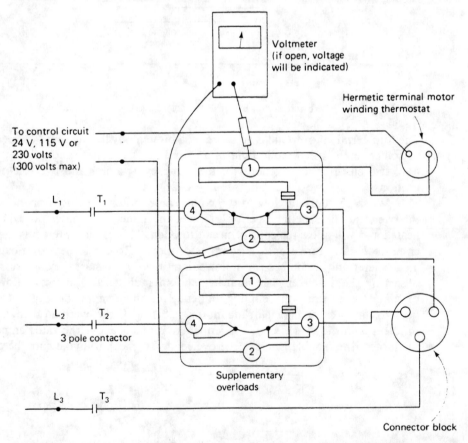

Figure 2-22. Checking voltage across control circuit terminals of a four-terminal overload.

pressor to be interrupted. When the hydraulic fluid has cooled sufficiently, the over-load may be manually reset and the compressor started again. To check these over-loads, place an ammeter on the wire to the compressor motor common terminal and start the compressor while observing the ammeter (see Fig. 2-23).

The ammeter should indicate a momentary current flow of approximately six times the rated running amperage of the compressor motor, then drop back to the rated amperage of the compressor motor or below. If the overload then cycles the compressor motor, the problem is in the overload. If the amperage draw remains above the rated amperage of the motor, the trouble is not in the overload. To check to make sure that the overload is cycling the compressor motor, check across the line terminals of the overload with a voltmeter while the compressor is off (see Fig. 2-24). If the overload is open, a voltage reading will be indicated. No voltage read-

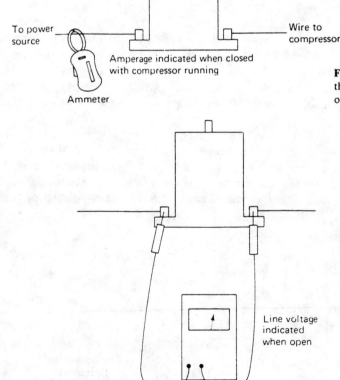

Figure 2-23. Checking amperage through a hydraulic-type compressor overload.

Figure 2-24. Checking voltage across a hydraulic-type compressor overload.

ing indicates that the overload has not opened the circuit. The trouble is elsewhere. These overloads are of the manual reset type. Be sure to replace them with an exact replacement for proper protection.

K. *Internal Motor Thermostats*

Internal motor thermostats function exactly like a two-terminal overload. They are wired in series with the compressor motor contactor coil and are placed inside the compressor housing and embedded in the motor winding to sense more accurately the temperature of the winding. They have two external line terminals on the outside of the housing and are generally connected to the control circuit. However, some manufacturers place them in the wire to the compressor motor common terminal (see Fig. 2-25).

To check these types of thermostats, turn off all electrical power and remove the wiring from the thermostat terminals. Then using an ohmmeter check across the thermostat terminals with the ohmmeter set on the R × 100 scale. Be sure to zero the ohmmeter. No continuity is an indication that the thermostat contacts are open. This is not, however, an indication that the thermostat is defective. It takes as long as 45 minutes for the thermostat to cool to its normal operating temperature. When it has cooled sufficiently, it will reset automatically. If the compressor housing feels hot to the touch, wait until it has cooled down so that your hand can be held on it comfortably. This is about 120°F. Then recheck the continuity of the thermostat contacts with an ohmmeter. In some instances it may be desirable to allow a trickle of water to flow over the compressor housing to aid in the cooling process. *Caution:* To prevent a possible electrical short, do not allow the water to enter the terminal box.

After the compressor has cooled sufficiently, and only as a last resort, the thermostat terminals can be jumpered. To prevent possibly burning out of the compressor motor, leave the jumper across the terminals for only a very short period of time. If the compressor should start, it can be reasonably assumed that the thermostat is defective, in which case the compressor will have to be replaced or the

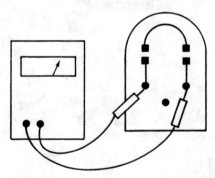

Figure 2-25. Checking internal compressor motor thermostat.

unit operated without this protection. In some cases the refrigerant returning to the compressor will help cool the thermostat enough so that the contacts will close. Check the contact continuity after the compressor has operated a few minutes. If it has reset, place the thermostat back in the circuit and check its operation. Permanent operation of the compressor with this thermostat jumpered is not recommended, but it can be done until a replacement compressor can be obtained.

L. Internal Compressor Motor Overloads

These overloads are sometimes used to provide the necessary protection. These controls are different from the internal thermostat because they are line-break devices which have no external wiring connections. The internal overload is embedded into the motor winding much the same as the thermostat, but it is wired in series with the compressor motor common terminal inside the compressor housing (see Fig. 2-26). Normally, compressors that are equipped with internal overload protection are marked near the terminal box to indicate their use.

To check the internal overload, remove all electrical leads from the compressor terminals. Then check the continuity between the compressor common and run terminals with the ohmmeter set on the R × 100 scale. Be sure to zero the ohmmeter. If no continuity is indicated, check between the start and run terminals for continuity. When no continuity is indicated at either of these points, the overload has tripped. The compressor must be allowed to cool down before the internal overload will reset automatically. A small trickle of water over the compressor housing will often help in cooling the compressor down. To prevent an electrical short, do not allow the water to enter the terminal box. If there is no continuity after the compressor has cooled down to the point where the hand can be held comfortably on the housing, the overload is defective and the compressor must be replaced before operation of the equipment can be resumed. In any case, make certain that the compressor has had sufficient time to cool before replacing it.

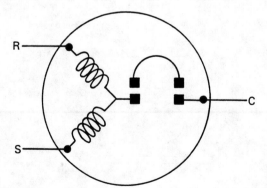

Figure 2-26. Internal overload wiring connections.

2-7 COMPRESSOR LUBRICATION

Heat pump compressors, like any moving mechanical device, require lubrication. The proper oil level should be maintained in the crankcase to provide this necessary lubrication. When the compressor is equipped with an oil sight glass, the oil level should be at or slightly above the center of the sight glass (see Fig. 2-27). On compressors not equipped with an oil sight glass, the manufacturer's recommendations should be followed. These recommendations will generally refer to the amount of oil in ounces for a given compressor model. When replenishing the oil in a hermetic compressor, it may be necessary to remove all of the oil from the housing and measure in the correct charge. The proper type of oil for the operating temperature should be used to ensure proper oil return and good compressor lubrication. Refrigeration oil containers must be kept sealed to prevent moisture and other contaminants from getting into the oil.

There are three types of lubrication systems used on air-conditioning and refrigeration compressors: splash, pressure, and a combination of both. The splash system is used on compressors up to 3 horsepower (hp). Compressors of 3 hp and larger are pressure lubricated. The oil pressure may be checked to determine if proper lubrication is being provided on forced lubrication compressors (see Fig. 2-28). A net oil pressure of 30 to 40 pounds per square inch (psi) [307.09 to 375.79 kilopascal (kPa)] is normal; however, adequate lubrication will be provided at pressures down

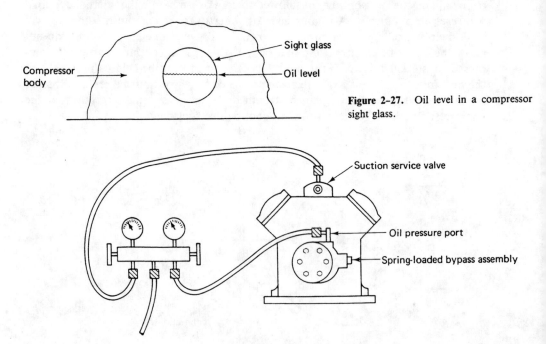

Figure 2-27. Oil level in a compressor sight glass.

Figure 2-28. Checking oil pressure.

to 10 psi (169.69 kPa). To obtain the net oil pressure, subtract the suction pressure from the oil pump pressure. Example: 90 psig pump pressure − 50 psig suction pressure = 40 psig net oil pressure (620.528 kPa − 344.738 kPa).

A. Compressor Bearings That Have Not Received Proper Lubrication

Bearings that have been subjected to this type of condition will become tight and sometimes will "freeze" the shaft and prevent operation. If this condition occurs, the compressor motor will become overloaded and will trip the overload or the circuit breaker, or will blow the electrical fuses. In any case, the amperage draw of the compressor motor will be higher than normal. When the compressor motor is not allowed to turn, it will draw locked rotor amperage. This amperage rating is indicated by "LR" on the motor nameplate. The compressor must be replaced when it is drawing the indicated LR amperage. The compressor oil should be checked and if it is found to be insufficient, the system must be checked to determine the reason for the lack of oil. The loss of oil may be due to refrigerant leaks, oil logging of the evaporator, low refrigerant charge, and so on. The reason must be corrected before the new compressor is placed in operation or it, too, may soon fail. Do not confuse this condition with faulty starting components or a faulty-running capacitor.

A malfunctioning oil pump will generally go undetected until the compressor bearings are damaged enough to cause knocking or until the compressor is frozen mechanically. A malfunctioning oil pump may be due to mechanical wear of the pump. It may become vapor-locked with refrigerant, or the inlet screen may become plugged with dirt or sludge. Should a worn pump be the culprit, it must be replaced at the same time that the compressor is being replaced or repaired. A vapor-locked oil pump will produce no oil pressure. The vapor lock must be removed by bleeding off the vapor through the gauge connection.

When the oil inlet screen becomes clogged, the oil flow will be restricted and perhaps completely stopped. In this case the screen must either be replaced or cleaned, along with a complete cleaning of the compressor crankcase, replacement of the oil charge, and replacement of the refrigerant filter-drier.

B. Compressor Bearings That Have Become Worn

Compressor bearings that become worn are usually noisy. The compressor loses its efficiency, and the result is poor refrigeration. Loose bearings are usually indicated by more noise than usual from the compressor. In some instances, the bearings will not knock but they will cause a vibration in the compressor. When the compressor is equipped with a forced lubrication system, the oil pressure will be lower than normal. The amperage draw will be from normal to low. In some cases the suction pressure may be high and the discharge pressure low. When this condition exists, the compressor must be replaced or overhauled, depending on the type. This condition is usually the result of age and usage and not faulty lubrication.

C. Compressor Valves

Compressor valves are the components that control the flow of refrigerant through the compressor. Should they become broken or start leaking, the compressor will become inefficient. A broken or leaking suction valve will result in a higher-than-normal suction pressure. To check for a defective compressor suction valve, connect the gauge manifold to the service valves on the compressor and open the service valves to the gauge port position (see Fig. 2–29).

Next, front seat the suction service valve (screw it all the way in), and observe the suction pressure while the compressor is operating. The suction pressure should pull down to at least 28 in. of vacuum (6.87 kPa) in 1 or 2 minutes. If it does not pump down to this reading, stop the compressor for 2 or 3 minutes; then restart it and allow it to run for 1 or 2 minutes. If the desired 28 in. of vacuum is not reached, the valves must be replaced. In the case of a hermetic compressor, the compressor must be replaced. When the fault is in open-type or semihermetic compressors, the compressor service valves must be closed, the refrigerant pressure in the compressor relieved, the head removed, and the valve plate replaced (see Fig. 2–30).

D. Broken or Leaking Discharge Valve

This condition will result in a lower-than-normal discharge pressure. To check for a defective compressor discharge valve, connect the gauge manifold to the service valves on the compressor and open the service valves to the gauge port connection (see Fig. 2–29). Next, front-seat the suction service valve (screw it all the way in), start the compressor, and allow it to pump as deep a vacuum as possible. Stop the compressor and observe the compound gauge. If the pressure rises, start the compressor and pump another vacuum. Stop the compressor and again observe the compound gauge. If the pressure increases again, close off the discharge service valve. If the pressure on the compound gauge stops rising, the discharge valve is

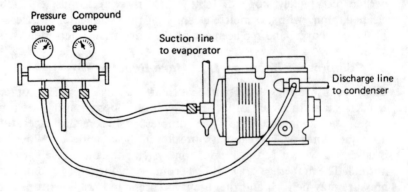

Figure 2–29. Gauges connected to compressor to take refrigerant pressure readings.

Figure 2-30.   Location of valve plate.

bad and must be replaced. In the case of a hermetic compressor, the compressor must be replaced. When the fault is in an open-type compressor, the compressor service valves must be closed, the refrigerant pressure in the compressor relieved, the head removed, and the valve plate replaced (see Fig. 2-30).

2-8 COMPRESSOR SLUGGING

Slugging is a noisy condition that occurs when the compressor is pumping oil or liquid refrigerant. When a compressor is slugging, it will sound like the clattering of an automobile engine that is under a strain. Continued slugging will probably result in broken valves, scored pistons, and galled bearings. It should be evident that a slugging condition must be corrected.

A. Oil Slugging of a Compressor

Oil slugging of a compressor occurs when there is too much oil in the compressor crankcase or when the compressor is started with liquid refrigerant in the crankcase. In this situation, some of the oil must be removed to maintain the oil level recommended by the compressor manufacturer. The excess oil may be drained through a drain plug, or some installations may require removal of the compressor so that the oil may be poured out. In either case the refrigerant must be pumped from the compressor or purged from the system. Do not attempt to remove oil from the compressor with refrigerant pressure in the crankcase.

To pump refrigerant from the compressor, connect the gauge manifold to the compressor service valves (see Fig. 2-29). Front-seat the suction service valve and allow the unit to run until approximately 2 psig (114.73 kPa) pressure is shown on the compound gauge. Do not pump the unit below atmospheric pressure for this procedure. Front-seat the compressor discharge service valve and relieve any remaining pressure through the gauge manifold. The compressor may now be serviced.

B. Refrigerant Slugging

The refrigerant slugging of a compressor is the result of liquid refrigerant being returned to the compressor. This can be caused by several things and will usually result in moisture condensing on the compressor housing because of the lower temperature, and sometimes ice or frost will form on the compressor housing. Some of the causes are: overcharge of refrigerant, especially on capillary tube systems; superheat setting too low on thermostatic expansion valve; automatic expansion valves open too much; or a low load condition on the evaporator. The obvious solution is to correct any of these causes. However, if these conditions cannot be remedied, a suction-line accumulator may be installed to prevent refrigerant slugging of the compressor (see Fig. 2–31). Almost all equipment manufacturers also install or recommend the installation of crankcase heaters to reduce slugging due to liquid refrigerant entering the crankcase.

2-9 OVERSIZED COMPRESSORS

Generally, an oversized compressor will produce unsatisfactory results. They will produce a lower-than-normal suction pressure, resulting in a lower-than-normal evaporator temperature. This lower-than-normal temperature causes excessive removal of humidity from the space. This moisture removal is critical and is not desirable in some industrial applications. The procedure is to replace the compressor with one of the proper size and capacity. However, in some cases it may be possible and less expensive to install a capacity reduction device on the compressor. The manufacturer of the compressor should be consulted to determine whether or not capacity reduction is satisfactory with the particular piece of equipment under question and its use. They can also give the proper specifications as to what type is best suited for that unit.

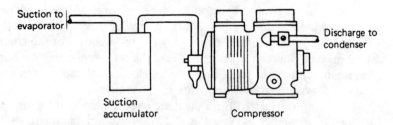

Figure 2–31. Suction-line accumulator installation.

2-10 *MOTOR STARTERS AND CONTACTORS*

The purpose of a motor starter or contactor is to provide the switching action of the high current and voltage required by the compressor. This is done by a signal given by the control circuit on demand from the thermostat or temperature controller. Motor starters and contactors are electromagnetic-operated devices.

A. *Burned Starter or Contactor Coil*

A burned starter or contactor coil will prevent operation of these devices because there will be no electromagnetic field to operate them. To check for a burned coil, turn off the electrical power to the unit and remove the electric wiring from the terminals of the coil. Check the continuity of the coil with an ohmmeter (see Fig. 2-32). If the coil is open, there should be no continuity indicated on the ohmmeter. Be sure to zero the ohmmeter before attempting to check continuity of any circuit. Many times the insulation on the coil will be discolored, indicating that it has been overheated. When there is no continuity indicated, the coil should be replaced with an exact replacement, or in some cases the entire starter or contactor must be replaced.

B. *Sticking Motor Starter or Contactor*

This condition can cause permanent damage to the motor. A starter or contactor that sticks may prevent the motor from starting or it may keep it running when there is no demand for it. When the starter or contactor sticks during the initial startup, it will usually buzz and either prevent starting of the motor or cause delayed starting of the motor. When the starter or contactor sticks closed, the compressor or motor will never stop. There are several types of sprays on the market that may be used to lubricate these troublesome and dangerous controls. However, it is generally recommended that starters and contactors that fall into this category be replaced.

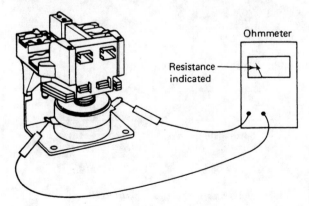

Ohmmeter

Resistance indicated

Figure 2-32. Checking continuity of contactor coil.

C. Burned Starter or Contactor Contacts

These faulty parts can cause permanent damage to the motor windings by preventing the proper flow of current to them. These contacts will be severely pitted and will not make good contact, thus causing a higher current draw than normal (see Fig. 2–33). In an emergency, these contacts may be lightly filed until the mating surfaces match (see Fig. 2–34). However, the damaged contacts should be replaced as soon as possible because they will again burn and become pitted in a very short time.

　　　Sometimes these contacts may become so bad that they will not make contact at all. This can be determined by energizing the starter or contactor and checking across the contact points with a voltmeter with the electric power on and the contactor closed by demand from the control circuit. To close the contacts manually may give a false reading (see Fig. 2–35). If the contacts are open, there will be line voltage indicated on the voltmeter. However, if the contacts are closed, there will be no indication of voltage at this point.

Figure 2-33.   Burned and pitted contacts.

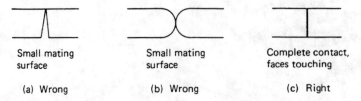

Small mating surface	Small mating surface	Complete contact, faces touching
(a) Wrong	(b) Wrong	(c) Right

Figure 2-34.   Right and wrong contact mating surfaces.

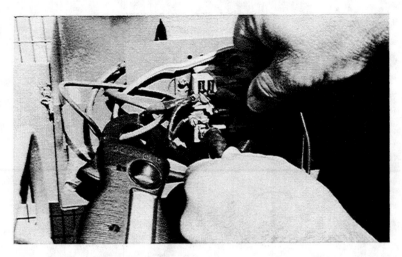

Figure 2-35. Checking voltage across contacts.

2-11 *CONTROL CIRCUITS*

The control circuit may be either electrical, electronic, or pneumatic. In the electric and electronic types, there is a small current flow through it. The pneumatic control circuit uses air pressure to operate the desired controls. These circuits usually contain various controls, such as the thermostat, contactor, starter, and various safety devices used to protect the compressors, motors, heat exchangers, and so on. To check the control circuit, make certain that the electric power to the unit is on and check the transformer or other source of control circuit power. If no power is present, there is a blown fuse, a tripped circuit breaker, or a defective transformer which must be replaced. Next, set the thermostat to call for the system to function, then check each component individually for voltage drop. If the control contacts are closed, no voltage will be indicated on the voltmeter. If the contacts are open, the applied voltage will be indicated (see Fig. 2-36). The control where the voltage reading is indicated must be checked further for possible replacement or repair.

2-12 *OIL FAILURE CONTROLS*

Oil failure controls are used to protect the compressor from a lack of lubrication. The control is actuated by the difference in pressure between the pump outlet and the crankcase pressure. A time-delay switch allows the oil pressure to build up to preset operating pressure on compressor startup and prevents nuisance shutdown of the compressor if the oil pressure drops for a short period of time.

To check for a faulty oil failure control, connect a compound gauge to the oil pump outlet. Be sure to leave the oil failure control connected. Connect a voltmeter across terminals 1 and 2 on the oil failure control. Start the compressor and observe the gauges and the voltmeter. The difference in pressure indicated on the two gauges should be at least 10 psi (169.69 kPa) in a short time. When this pressure differential

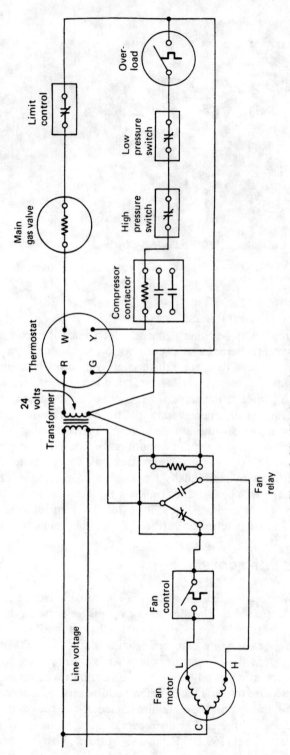

Figure 2-36. Heating and cooling control circuit.

is reached, the contacts between terminals 1 and 2 should open and a voltage indicated on the voltmeter (see Fig. 2-37). However, if this minimum pressure is reached and the contacts remain closed, the control will not stop the compressor in about 2 minutes. The control is faulty and must be replaced.

The time delay of these controls is based on 120 or 240 volts (V) as applied in an ambient temperature of 75°F (23.9°C) with the cover in place. If the ambient temperature is much higher than 75°F (23.9°C), the control may be causing nuisance shutdown because of a high ambient temperature rather than low oil pressure. In this case the control must be relocated to a cooler ambient temperature.

2-13 THERMOSTATS

Thermostats are temperature-sensitive devices used to control equipment in response to the demands of the space in which they are located. Thermostats may be operated by a bimetal or a "feeler" bulb filled with a fluid that expands and contracts in response to temperature changes.

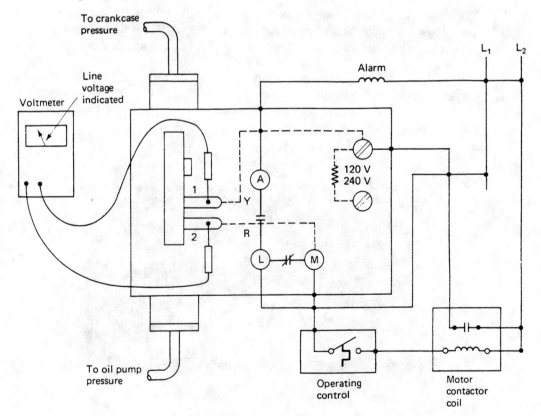

Figure 2-37. Checking operation of oil safety control.

A. Room Thermostat

A room thermostat generally makes use of a bimetal element for its operation. This type of thermostat is the most popular for heating installations (see Fig. 2–38). To check the thermostat turn it above room temperature and place a reliable thermometer as close as possible to the bimetal element. Allow the thermometer to remain there for 10 minutes. Next turn the thermostat temperature selector down. The contacts should "make" (close) at no more than $2°F$ $(1.11°C)$ below the temperature indicated by the thermometer. If not, the thermostat must be calibrated. There are several means of calibrating a thermostat. Therefore, the manufacturer's specifications should be consulted. If more than $10°F$ $(5.56°C)$ calibration is needed, replace the thermostat. To check for an inoperative thermostat, check the voltage from the red terminal to the heating terminal. If the contacts are closed, no voltage will be indicated. If the contacts are open, a voltage will be indicated.

B. Heating Anticipator

The heating anticipator incorporated in a thermostat is used during the heating cycle only. Most heat anticipators are adjustable (see Fig. 2–39). When adjusting a heat anticipator, the total amperage draw of the control circuit must be known. To determine the amperage draw, wrap the tong of an ammeter with one wire to the main gas valve, or the compressor or heating element contactor coil and read the amperage draw while the circuit is in operation (see Fig. 2–40). Divide the amperage indicated by the number of turns taken on the ammeter tong. Set the heat anticipator to match this amperage draw.

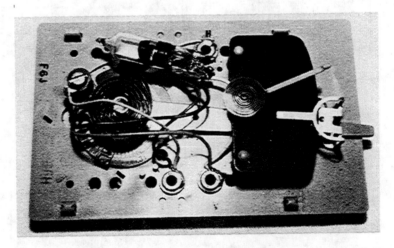

Figure 2–38. Thermostat with cover removed.

Figure 2–39. Heating anticipator.

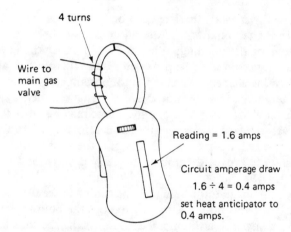

4 turns

Wire to
main gas
valve

Reading = 1.6 amps

Circuit amperage draw

1.6 ÷ 4 = 0.4 amps

set heat anticipator to
0.4 amps.

Figure 2–40. Checking amperage in a temperature control circuit.

C. Location of the Thermostat

The thermostat location is very important to satisfactory operation of the equipment. The thermostat should be located on an inside wall about 5 ft (1.52 m) from the floor. It should not be affected by any external heat source, such as lights, sun, television, and so on. It should be located so that it will sense the average return air temperature.

D. Thermostat Switches

These switches are placed according to the type of operation desired. The system switch controls operation of the equipment. The fan switch controls the operation of the fan. Both switches are usually incorporated in the thermostat subbase (see

Figure 2–41. Air-conditioning thermostat.

Fig. 2–41). To check out the switches, use a jumper wire to jump from the "R" or "V" terminal to the "W" or "H" terminal for heating. The first letter of each of these sets refers to Honeywell thermostats; the second refers to General Controls thermostats. When the fan switch is set to "on," the fan will run continuously. The user can select the operation desired. Should these switches become defective, the subbase or thermostat must be replaced before normal operation can be resumed.

E. Outdoor Thermostats

These are remote-bulb-type thermostats that are used on heat pump systems. They are sometimes mounted in the terminal box of the outdoor unit. Other times they are mounted under the eave of the building. When mounted under the eave, protection from the wind, rain, and sun must be provided. These thermostats are used to energize the auxiliary heating elements when the outdoor temperature falls below the balance point of the building. The thermostat setting is determined by the designer to provide the greatest efficiency and economy. There may be more than one outdoor thermostat. Therefore, each one is set at a temperature equal to the combined output balance point of all the other heat strips plus the heat pump. Electric power for these thermostats is provided through the second stage of the indoor thermostat (see Fig. 2–42).

To check the outdoor thermostat, the bulb must be cooled to see if the contacts open and close at the desired temperature. One way to cool the bulb is to insert it in an ice and salt solution along with a thermometer. Adjust the thermostat to the desired temperature if possible. If adjustment is not possible, replace the thermostat.

Figure 2–42. Outdoor thermostat.

2-14 *PRESSURE CONTROLS*

Pressure controls are designed to protect compressors and motors from damage as a result of excessive pressures (see Fig. 2-43). Low-pressure controls are used to open the control circuit when the refrigerant pressure in the low side of the system falls below a given pressure. High-pressure controls are used to open the control circuit when the refrigerant pressure in the high side of the system rises to a given pressure. These pressure settings are generally recommended by the equipment manufacturer. Usually, when these controls cause the compressor to cycle, the problem is due to some cause other than the control.

A. *Low-Pressure Controls*

These controls are designed to respond to the refrigerant pressure in the low side of the system. They will stop the compressor motor, to protect it from overheating and to prevent the compressor from pumping oil out of the crankcase. On some smaller units the low-pressure control may also be used as a temperature control. To check the low-pressure control, install a compound gauge on the compressor suction service valve (see Fig. 2-44).

Figure 2-43.   Dual pressure control. (Courtesy of Control Products Division, Johnson Controls, Inc.)

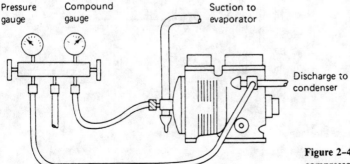

Figure 2-44.   Gauges installed on a compressor.

Do not disconnect the low-pressure control or cause it to become inoperative. Crack the service valve off the back seat. With the compressor running, front-seat the suction service valve and observe the pressure on the compound gauge when the pressure control stops the compressor. If the actual pressure does not correspond with the desired control setting, adjust the control, back-seat the suction service valve, and repeat the foregoing procedure until the desired cut-out and cut-in points are obtained. Rarely do these controls need replacement except in the case of refrigerant leakage, in which case replacement is preferred to repair.

B. High-Pressure Controls

High-pressure controls are designed to respond to the pressure in the high side of the system. They will stop the compressor motor to protect it from being overloaded due to excessively high discharge pressures. To check a high-pressure control, install a pressure gauge on the compressor discharge service valve. Crack the service valve off the back seat. Block the airflow through the condenser, or stop the water pump if a water-cooled unit is used (see Fig. 2–45). Start the compressor and observe the pressure on the pressure gauge when the control stops the compressor. If the actual pressure does not correspond with the desired setting, adjust the control, push the reset, and repeat the procedure above until the desired cutout point is obtained. Rarely do these controls need to be replaced except in the case of refrigerant leakage, in which case, replacement is preferred to repair.

2–15 LOOSE ELECTRICAL WIRING

Loose electrical wiring can cause many problems and at times be extremely difficult to locate. The problems caused by loose electrical wiring do not follow any set pattern and are usually difficult to find. Most loose wiring can be found by visual

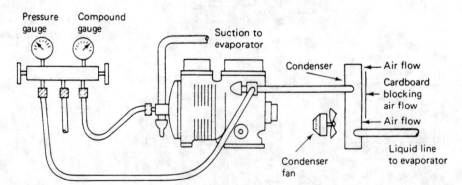

Figure 2-45.   Blocking airflow to raise discharge pressure and checking high-pressure control.

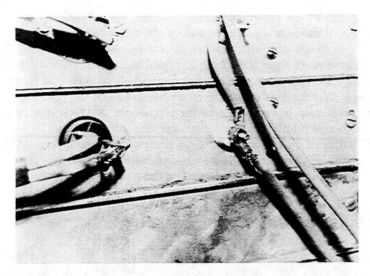

Figure 2-46. Hot electrical joint.

inspection (see Fig. 2-46). However, when a loose electrical wire is suspected, it is at times necessary to check each wire and its connections individually. This is usually time-consuming and a great task. But it must be done before the unit will operate satisfactorily again. Once the bad wire or connection is found, it must be replaced or repaired. Any wiring that is not properly repaired will only cause problems in the future.

2-16 IMPROPERLY WIRED UNITS

Improperly wired units will operate inefficiently, if at all. They will not produce the results desired of the equipment. Each manufacturer designs its own wiring diagrams for each piece of equipment, designs that will cause the unit to produce the desired results. If a service technician is in doubt about the wiring on a piece of equipment, the recommended wiring diagram should be consulted and the wiring changed to match the diagram.

2-17 LOW VOLTAGE

Low voltage to a motor can cause it to overheat and will do damage to the motor windings. This overheating is due to excessive current draw because of the low voltage. This does not include those types of motors that are designed to operate on a varied voltage, such as some fan motors. There are several causes of low voltage, such as wire too small, loose connections, bad contactor contacts, or low voltage provided by the power company. To check for low voltage, connect the voltmeter to the common and run terminals of the motor (see Fig. 2–47).

Start the unit and observe the voltage indicated on the voltmeter. It should not vary more than 10% from the rated voltage of the unit. If a voltage drop of more than 10% is detected, check the size of the wire leading to the unit. Be sure that it is at least as big as the recommendation of the equipment manufacturer. If not, it should be replaced with the proper size. If the wire is of sufficient size and the voltage is still low, check for loose connections. These connections will usually be indicated by the wire insulation being overheated or burned. Repair these connections. If there are no loose connections, check the voltage at the electric meter. If it is found to be low here, contact the power company for assistance.

2–18 STARTING AND RUNNING CAPACITORS

Capacitors are used by many manufacturers to improve the starting and running characteristics of their motors. Capacitors are manufactured for use in both starting and running cycles of electric motors. Each manufacturer determines the proper size for its motor. These recommendations should be followed.

A. Starting Capacitors

Starting capacitors are used in the starting circuit of the motor. They are generally round, encased in a plastic casing, and have a relatively high microfarad (μF) rating (see Fig. 2–48). Starting capacitors are designed for short periods of use only. Pro-

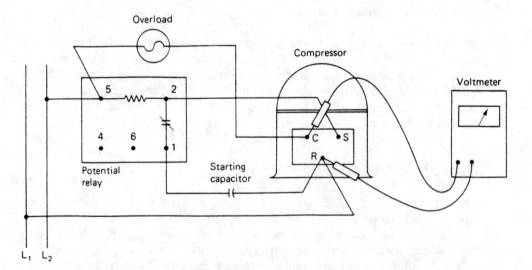

Figure 2–47. Checking starting and running voltage to a motor.

Figure 2-48. Starting capacitor with bleed resistor.

longed use will generally result in damage to the capacitor. If a starting capacitor is found to be defective, be sure to check the starting relay before the unit is placed back into service or the new capacitor may also be damaged. The best way to check a capacitor is to use a capacitor analyzer. This type of meter provides a direct reading of the microfarad output of the capacitor without the use of bulky equipment and mathematical formulas (see Fig. 2-49). Starting capacitors that are found to be out of the range of 0 to +20% of the microfarad rating of the capacitor must be replaced with the proper size.

Replacement capacitors must have at least the same or a greater voltage rating as those being replaced. When replacing starting capacitors, be sure that a 15,000 to 18,000-ohm (Ω) 2-watt (W) resistor is soldered across the terminals of the ca-

Figure 2-49. Capacitor analyzer. (Courtesy of Robinar Division-Sealed Power Corporation.)

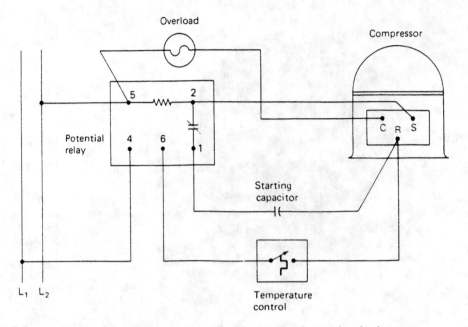

Figure 2-50. Compressor wiring diagram showing starting circuit.

pacitor (see Fig. 2–48). This is a precautionary measure to prevent arcing and burning of the starting relay contacts. To wire a starting capacitor into the circuit, see Fig. 2–50.

B. Running Capacitors

Running capacitors are in the operating circuit continuously. They are normally oil-filled capacitors. The microfarad rating of these capacitors is relatively low, even though they are physically large in size. Running capacitors are used to improve the running efficiency of motors. They also provide enough torque to start PSC (permanent split capacitor)-type motors. Running capacitors are provided with a red dot on one terminal (see Fig. 2–51). The red dot indicates the terminal that is most likely to short out in case of a capacitor breakdown. Because of the relatively high voltage generated in the start winding, the unmarked terminal is connected to the starting terminal on the motor. If the terminal with the red dot is connected to the motor starting terminal, damage to the winding could occur if the capacitor should short out. The best way to check these capacitors is with a capacitor analyzer. This type of meter provides a direct reading of the microfarad output of the capacitor without the use of bulky equipment and mathematical formulas (see Fig. 2–49). Running capacitors that are found to be out of the range of + (plus) or − (minus) 10% of the microfarad rating of the capacitor must be replaced with the proper size. The voltage rating of any capacitor must be equal to or greater than the one being replaced. A higher-than-normal running amperage usually indicates a weak capacitor. To wire a running capacitor into the circuit (see Fig. 2–52).

Figure 2–51. Run capacitor.

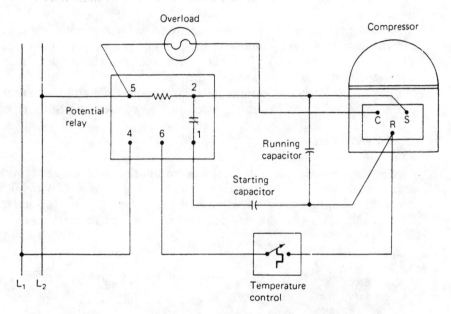

Figure 2–52. Compressor wiring diagram using start and run capacitors.

2-19 STARTING RELAYS

Starting relays are the devices used to remove the starting circuit from operation when the motor reaches approximately 75% of its normal running speed. Its function is basically the same as the centrifugal switch used in split-phase motors. There are four types of starting relays in use today. They are (1) amperage (current) relay, (2) hot wire relay, (3) solid-state relay, and (4) potential (voltage) relay. There is also a solid-state starting device which is used on PSC motor-compressors. The horsepower size and the design of the equipment regulate which type of starting relay is used.

63

A. Amperage (Current) Relay

This is an electromagnetic-type relay, which is normally used in ½-hp units and smaller (see Fig. 2–53). These relays are positional types and must be properly mounted for satisfactory operation. They must be sized for each motor horsepower and amperage rating. To check an amperage relay, turn off the electricity to the unit and remove the wire from the "S" terminal and touch it to the "L" terminal on the relay (see Fig. 2–54). Place an ammeter on the common wire to the compressor. Start the compressor and immediately remove the "S" wire from the "L" terminal. If the compressor continues to run and the amperage draw is within the amperage rating of the compressor, replace the relay. *Caution:* To prevent electrical shock, do not allow the loose end of the wire to come into contact with anything or anyone.

An amperage relay that is too large for a motor may not allow the relay contacts to close, thus leaving out the starting circuit. The motor probably will not start under these conditions. A relay that is rated too small for a motor may keep the contacts closed at all times while electrical power is applied to the unit, leaving the starting circuit energized continuously. Damage to the starting circuit may occur under these conditions. A motor protector must be used with this type of relay.

B. Hot Wire Relay

Hot wire relays are a form of current relay, but they do not operate with an electromagnetic coil. They are designed to sense the heat produced by the flow of electrical current through a resistance wire (see Fig. 2–55). There are two sets of contacts in these relays, a set for starting and a set for running. To check these relays, turn

Figure 2–53. Amperage relay.

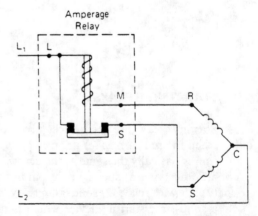

Figure 2–54. Amperage relay connections.

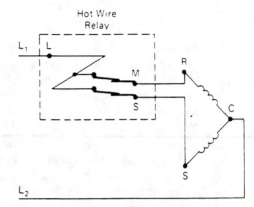

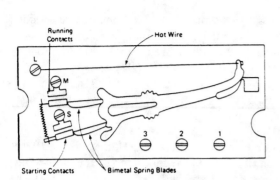

Figure 2-55. Hot wire relay.

Figure 2-56. Hot wire relay connections.

off the electrical power to the unit and remove the wire from the "S" terminal on the relay and touch it to the "L" terminal on the relay (see Fig. 2-56).

Place an ammeter on the common wire to the compressor. Start the compressor and immediately remove the "S" wire from the "L" terminal. To prevent electrical shock, do not allow the loose end of the wire to touch anything or anyone. If the compressor continues to run and the amperage draw is within the amperage rating of the compressor, replace the relay. If the compressor operates within the amperage rating indicated by the manufacturer but still stops within 1 or 2 minutes, the overload portion of the relay is defective. Replace the relay with one of the proper size for the horsepower and amperage rating of the compressor motor.

A hot wire relay that is too large for the motor will not remove the starting components from the circuit, resulting in possible motor damage. One that is rated too small will stop the motor with the overload after 1 or 2 minutes of operation. An additional overload is not necessary when these relays are used. These are non-positional relays.

C. Solid-State Starting Relay

These types of relays use a self-regulating conductive ceramic which increases in electrical resistance as the compressor starts, thus quickly reducing the starting winding current to a milliampere level. The relay switches in less than 0.35 second when the current flow is at least 10 amperes (A). This allows this type of relay to be applied to refrigerator compressors without being tailored to each particular system within the specialized current limitations. These relays will start virtually all split-phase 115-V hermetic compressors up to $\frac{1}{3}$ hp. An overload must be used with these relays. Since these relays are push-on devices, the easiest method of checking their operation is to install a new one. Be sure to check the amperage draw of the compressor motor.

D. Potential Relay

Potential relays operate on the electromagnetic principle. They incorporate a coil of very fine wire wound around a core. These starting relays are used on motors of almost any size. They are nonpositional. The contacts are normally closed and are caused to open when a plunger on the relay is pulled into the relay coil. These relays have three connections to the inside in order for the relay to perform its function. These terminals are numbered 1, 2, and 5. Other terminals, numbered 4 and 6, are sometimes used as auxiliary terminals (see Fig. 2-57).

   To check a potential relay, turn off the electrical power to the unit and remove the wire from terminal 2 on the relay and touch it to terminal 1 on the relay. Place an ammeter on the wire to the common terminal on the motor. Start the motor and immediately remove the wire 2 from terminal 1 on the relay. *Caution:* To prevent electrical shock, do not allow the loose wire to touch anything or anyone. If the compressor continues to run and the amperage draw is within the rating of the compressor, replace the relay.

   The sizing of potential relays is not as critical as with the amperage and hot wire relays. A good way to determine what relay is required is to start the motor manually and check the voltage between the start and common terminals on the compressor while the motor is operating at full speed (see Fig. 2-58). Multiply this voltage by 0.75 and this will be the pickup voltage of the relay.

E. Starting Relay Mounted So That It Vibrates

A starting relay that is mounted in this type of location will cause the contacts to arc excessively and become burned. When a relay is mounted on such a surface, it must be remounted on a more solid surface. Generally, the relay will need to be

Figure 2-57. Potential starting relay.

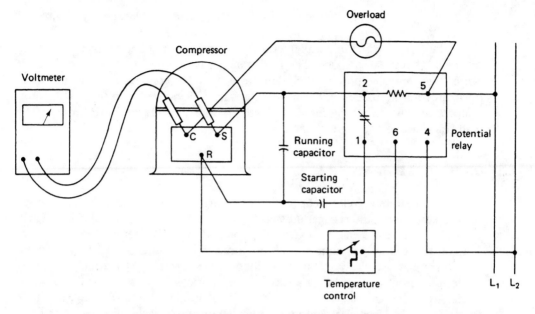

Figure 2–58. Checking voltage between start and common terminals.

replaced during the process. Be sure to replace the relay with the proper size and type, and if it is a positional relay, it must be mounted to satisfy the manufacturer's recommendations.

F. Positive Temperature Coefficient Starting Device

These PTC resistor devices are used on PSC motors to provide additional starting torque. Their use is not recommended on systems that use a thermostatic expansion valve as the flow control device or on systems that are subjected to short cycling conditions. When they can be used, however, they are very simple to install, are less expensive to buy, and have a wide range of application (see Fig. 2–59). The PTC material has a steep slope positive temperature coefficient which has a cold resistance rating of about 50 Ω and a hot resistance rating of about 80,000 Ω.

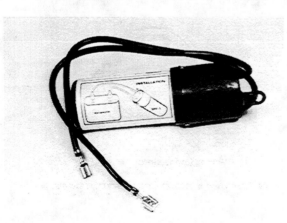

Figure 2–59. Positive temperature coefficient starting device.

67

The PTC is wired in parallel with the run capacitor and increases the starting torque to about 200 to 300%. The PTC material heats up and takes the start assist out of the circuit in approximately $\frac{1}{5}$ of a second. The compressor then runs in its normal operating mode. To check this start-assist device, allow it to cool to room temperature and check the resistance with an ohmmeter. If it is very far from the cold resistance rating, replace the device.

2-20 CRANKCASE HEATERS

Crankcase heaters are electrical resistors that are designed to provide just enough heat to keep any liquid refrigerant that might enter the crankcase boiled off. There are two different designs: (1) externally mounted, and (2) internally mounted (see Fig. 2-60). Externally mounted crankcase heaters are manufactured by several man-ufacturers and are easily installed on most compressors. The internally mounted heaters are usually designed for a specific model of compressor. To check crankcase heaters, dampen the finger and gently touch the heater. If it is working, it will be hot to the touch. The electrical power must be on for several hours before this test can be performed. Another method of checking is to disconnect both electrical wires and check the resistance of the heater (see Fig. 2-61). Crankcase heaters are gen-erally energized continuously and are designed to prevent overheating of the com-pressor oil.

2-21 TWO-SPEED OUTDOOR FAN THERMOSTAT

Two-speed outdoor fan motors are controlled by a thermostat mounted on a return bend of the outdoor coil. They sense both the outdoor temperature and refrigerant temperatures and automatically change the speed of the fan. These controls are

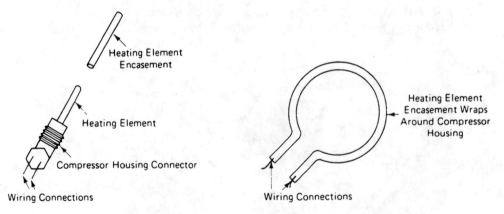

Figure 2-60.   (a) Internal crankcase heater; (b) external crankcase heater.

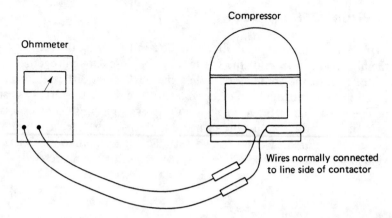

Figure 2-61.   Checking resistance of crankcase heater.

nonadjustable and switch the fan to low speed when the outdoor temperature is about 75°F (23.8°C) and the refrigerant temperature is about 95°F (35.0°C). The fan is switched into high speed when the outdoor temperature reaches about 90°F (32.1°C) and the refrigerant temperature is about 110°F (43.3°C) (see Fig. 2-62). To check these controls, check both the outdoor and refrigerant temperatures and if the fan is not operating in the proper mode, and electricity is not directed to the proper circuit, the thermostat is defective and must be replaced. Be certain that electricity is supplied to the unit.

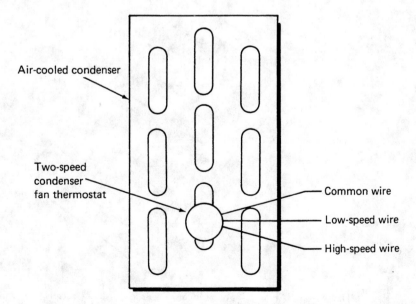

Figure 2-62.   Location of two-speed condenser fan thermostat.

2-22 UNEQUALIZED PRESSURES

"Unequalized pressures" is a term used to indicate that the refrigerant pressures in the high and low sides of the system are not close enough together to allow a PSC motor to start the compressor. When there is a great difference between these two pressures, the starting load is too great for the motor to start. Less starting torque is required as the difference between these two pressures becomes less; that is, with zero pressure differential between the high and low side, a minimum of starting torque is required. There are two remedies to this situation. One is to allow the system to sit idle for a longer period of time using a time-delay device (see Section 2–23). The second is to install a hard-start kit (see Section 2–24).

2-23 COMPRESSOR TIMER

A compressor timer is a device used to prevent the compressor from cycling too soon after shutdown. This control prevents starting of the compressor until the pressures have equalized to a safe starting point. This procedure helps to prolong the life of the compressor (see Fig. 2–63).

At the end of an on cycle the timer goes through a brief time period when the compressor cannot restart, usually 3 to 5 minutes. After the time period has passed the compressor can restart and operate normally. In most cases the timer is used on these units which do not have a hard-start kit installed.

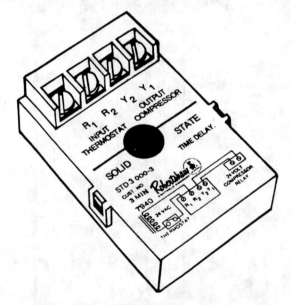

Figure 2–63.   Compressor timer. (Courtesy of Robertshaw Controls Co., Uni-Line Division.)

To check out this control, after the proper time period has passed with the system demanding operation, jumper the terminals R1 to Y1 and R2 to Y2. If the timer is defective, the compressor should start. If it does, replace the timer with a proper replacement.

2-24 HARD-START KIT

Hard-start kits are designed for use when conditions are encountered that prevent the PSC motor from starting normally. Such conditions are when the electrical power fluctuates or is too low to provide the necessary power for the proper starting of a compressor motor. Another such condition occurs when a PSC motor is used on a system that requires rapid cycle operation and less pressure equalization. Hard-start kits are designed to convert PSC motors to CSCR (capacitor start, capacitor run) motors. Hard-start kits consist of the proper starting relay and the proper starting capacitor, together with the necessary wiring to install the kit on the unit. The individual components may also be combined to make up a hard-start kit. To install a hard-start kit, turn off the electric power to the unit and complete the connections as shown in Fig. 2–64.

2-25 HIGH DISCHARGE PRESSURE

A high discharge pressure can be the result of one or a combination of several things. It is a condition that causes an overload on the motor and decreases the efficiency of the compressor and the refrigeration system. The most common causes

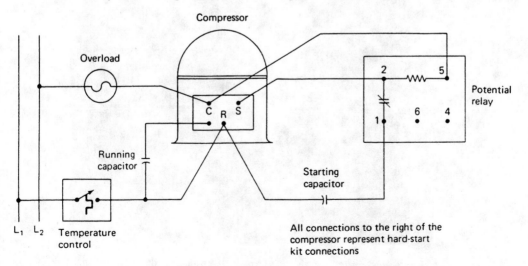

Figure 2-64. Diagram showing hard-start kit.

are (1) compressor discharge service valve closed, (2) lack of cooling air, (3) lack of cooling water, (4) overcharge of refrigerant, (5) noncondensable gases, or (6) a combination of these things.

A. Closed Compressor Discharge Service Valve

When front-seated, the compressor discharge valve will reduce or completely stop the flow of refrigerant from the compressor. Caution should be exercised to prevent this condition because damage to the compressor or motor is likely. The rapid buildup of pressure within the cylinder head is tremendous and increases as the piston completes its upward stroke. Never front-seat the compressor discharge valve while the compressor is running or start the compressor with the valve front-seated.

B. Lack of Cooling Air

The lack of cooling air over an air-cooled condensing coil will cause the discharge pressure to increase. This is because a higher pressure is required to condense the vapor to a liquid. The higher refrigerant temperature also reduces the unit efficiency because of the increased flash gas, as well as reducing the compressor efficiency. This condition is generally caused by a dirty condensing coil, loose fan belt, or bad fan motor bearings.

A dirty condensing coil can be cleaned by using a garden hose with a high-pressure nozzle. The water should be forced through the coil from both sides. Be sure to prevent water from entering the fan motor, which might cause an electrical short. This can be done by wrapping a piece of sheet plastic around the motor. Be

Figure 2-65. Cleaning air-cooled condenser with a garden hose and pressure nozzle.

sure to remove the plastic before starting the unit. Turn the unit off during this procedure (see Fig. 2-65).

A loose or broken fan belt will prevent the blower moving air to cool the coil. This condition is generally obvious and is easily corrected. To adjust the belt tension, turn the adjustment until the belt can be flexed about 1 in. with one finger using moderate pressure (see Fig. 2-66). If the belt is frayed, has wear grooves on the sides, or has become hard, it should be replaced. Be sure to use the proper-size belt. A belt that is too narrow will ride the bottom of the pulley (see Fig. 2-67). It will slip, causing decreased efficiency. A belt that is too wide will ride high in the pulley, will not maintain the desired efficiency, and will possibly overload the fan motor.

Bad fan motor bearings will cause the motor to overheat and cut out on the overload. When the condensing coil fan motor stops, the coil overheats and the compressor will cut out on high head pressure. To check for bad bearings, stop the unit, remove the belt, and move the motor shaft from side to side, *not* from end to end. Any movement of the shaft from side to side indicates bad bearings. The bearings must be replaced or the motor replaced, depending on the motor size.

C. Lack of Cooling Water

The lack of cooling water on a water-cooled condensing coil will cause the discharge pressure to increase because of the higher pressure required to condense the vapor

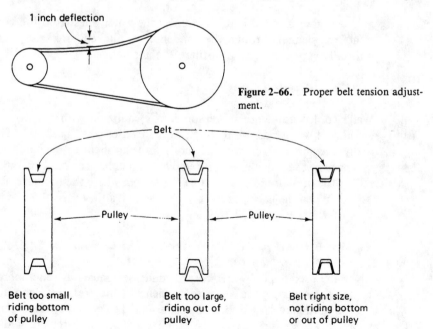

1 inch deflection

Figure 2-66.   Proper belt tension adjustment.

Belt

Pulley — Pulley

Belt too small,
riding bottom
of pulley

Belt too large,
riding out of
pulley

Belt right size,
not riding bottom
or out of pulley

Figure 2-67.   Comparison of belt and pulley fitting.

to a liquid. The higher refrigerant temperature also reduces the unit efficiency because of the increased amount of flash gas, as well as reducing the compressor efficiency. This condition is generally caused by plugged water stainers, pumps, or spray nozzles.

When this condition is suspected, check the temperature rise of the water as it passes through the condensing coil. This temperature rise should not be more than 10°F (5.56°C). If the temperature rise is greater than 10°F, a strainer, pump, or spray nozzle is stopped up or there is not enough water in the cooling tower sump. The necessary steps must be taken to relieve this condition.

A temperature rise of less than 10°F (5.56°C) indicates that the condenser is scaled and must be cleaned. There are several commercial cleaners available for cleaning (acidizing) water-cooled condensers. The amount used is determined by the manufacturer of the cleaner. Caution should be exercised to prevent damage to the equipment, personnel, and the surrounding vegetation when acidizing a unit. The most common method of acidizing a unit is to be sure that the strainers, pump, and spray nozzles are clean. Then dump the cleaner into the tower sump and check the mixture with pH strips, adding cleaner until the desired pH is indicated, and allow the unit to run until all the scale has been removed. The cleaner and sediment must be completely removed from the system. To do this, drain and flush the tower, condenser, and water lines with fresh water. Then add a neutralizer, allowing the system to run until the neutralizer has had time to neutralize the cleaner; then drain and refill the system with fresh water. Never leave the cleaner in the system because of possible damage to the equipment.

Another method is to disconnect the water lines from the condenser and circulate the cleaner through the condenser with a separate acid pump. This method is usually very expensive and therefore not popular.

D. Water-Cooled Condensers

Water-cooled condensers are sometimes equipped with a valve that controls the amount of water passing through the condenser (see Fig. 2–68). These valves eventually become clogged with minerals, causing them to become inoperative. When this occurs, the valve must be removed and either replaced or repaired. Be sure to relieve the pressure in that part of the system where the pressure attachment is connected. When the valve is repaired or replaced, adjust the valve to provide the proper flow of water (see Fig. 2–69).

E. Water Used to Cool a Water-Cooled Condenser

The water used to cool water-cooled condensers must have a temperature low enough to cause the refrigerant vapor to condense at the pressures encountered in the normal operating cycle. Cooling towers are used to cool the water (see Fig. 2–70). The spray nozzles cause the water to atomize and mix with the air. This mixing with the air causes some of the water to evaporate, lowering its temperature. If the spray

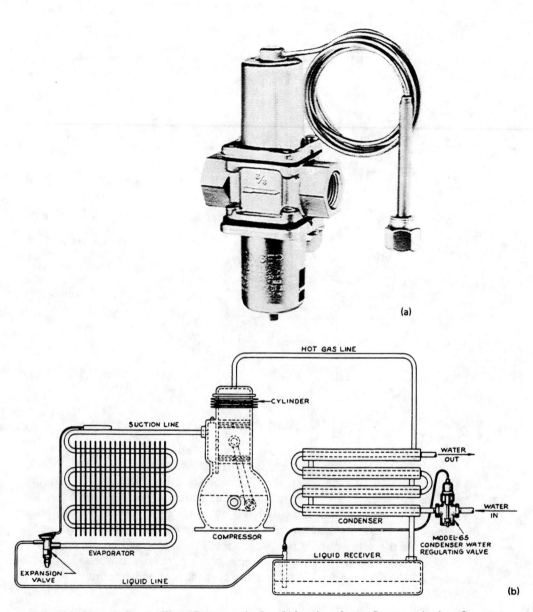

(a)

(b)

Figure 2-68.   (a) Water flow control valve; (b) location of water flow control valve. (Courtesy of Conrols Company of America.)

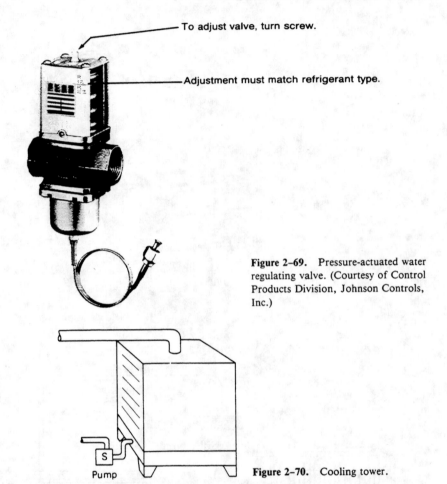

To adjust valve, turn screw.

Adjustment must match refrigerant type.

Figure 2–69. Pressure-actuated water regulating valve. (Courtesy of Control Products Division, Johnson Controls, Inc.)

Pump

Figure 2–70. Cooling tower.

nozzles do not atomize the water, it will not be cooled sufficiently. To determine whether or not the cooling tower is functioning properly, measure the water temperature as it leaves the tower sump and as it leaves the spray nozzle. The difference in these two temperature readings should be a minimum of 10°F (5.56°C) (see Fig. 2–71). Should the spray nozzles become clogged, the water will not be cooled sufficiently. Be sure that a full cone of water is coming from each nozzle. It may be necessary to clean the pump and strainers so that the required amount of water can be circulated through the system.

F. Overcharge of Refrigerant

An overcharge of refrigerant will cause a high discharge pressure because the extra refrigerant will take up space in the condenser that is needed to condense the vapor to a liquid. An overcharge of refrigerant will cause at least one-half the condensing

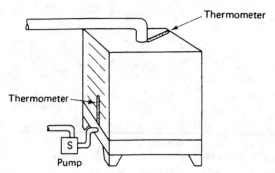

Figure 2-71. Checking cooling-tower water temperature.

coil tubes to be cooler than the remainder. The cool tubes are the ones that are full of liquid refrigerant (see Fig. 2-72). When a capillary tube flow control device is used, the suction pressure will also be higher than normal. The suction line will be cooler than normal and may frost over, depending on the amount of overcharge.

The excess of refrigerant must be purged from the system. When purging, allow only a small amount of refrigerant to escape at a time. This procedure is recommended to prevent purging too much refrigerant from the system, which would require that refrigerant be added back into the unit. Also, it prevents the removal of lubricating oil from the compressor crankcase.

G. Presence of Noncondensables

The presence of noncondensables in the system will cause a high discharge pressure because they will not condense under the normal operating pressures encountered in refrigeration systems and will take up space needed by the refrigerant in the condensing coil. To determine if noncondensables are present in a system, pump the

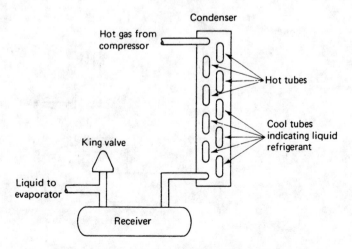

Figure 2-72. Checking liquid level in condenser.

system down; that is, pump all the refrigerant into the receiver or condenser by front-seating the liquid-line service (king) valve (see Fig. 2–73).

Operate the compressor until the low side has been pumped down to approximately 5 psig (235.34 kPa). Stop the compressor and allow it to stand idle until it has cooled to ambient temperature. Compare the idle discharge pressure to the pressure indicated on a pressure–temperature chart at that ambient temperature. If the discharge pressure is higher than that indicated by the chart, slowly purge the noncondensables from the highest point in the system. Purge the air slowly to prevent purging an excessive amount of refrigerant from the system. If the system contains only a small charge of refrigerant, it will probably be better to purge the entire charge, evacuate the system, and add a complete new charge. Noncondensables should be kept out of the system because they not only reduce the system efficiency but also contain moisture, which is extremely harmful to the refrigeration system.

2–26 LOW SUCTION PRESSURE

A suction pressure that is lower than normal may be due to any or all of the following reasons: (1) shortage of refrigerant, (2) dirty air filter, (3) dirty evaporator, (4) dirty blower, (5) loose fan belt, (6) iced-over evaporating coil, (7) TXV (thermostatic expansion valve) superheat set too low, or (8) restriction in the refrigerant system. The problem or problems causing low suction pressure should be found and corrected because the compressor may pump all the lubricating oil out of its crankcase and into the system, resulting in possible damage to the compressor as well as inefficient operation of the unit due to an oil-logged evaporating coil.

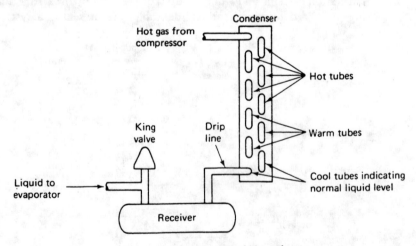

Figure 2–73. Location of king valve.

A. *Shortage of Refrigerant*

A shortage of refrigerant is usually due to a leak that has developed in the refrigerant circuit. The leak should be found and repaired and the system recharged with a proper charge of refrigerant. If the leak is not found and repaired, the refrigerant will escape from the system, requiring more service. A refrigerant leak is generally indicated by oil on the place where the leak has occurred. Many times, visual inspection will locate the problem area (see Fig. 2–74).

If the problem area is not found by visual inspection, an electronic or halide-torch leak detector should be used. Either of these will pinpoint a leak under most conditions. Two conditions that make leak detection difficult with these tools are when the wind is blowing outdoors or in an enclosed space where the refrigerant concentration is high. When these conditions are encountered, a soap-and-water solution or one of the liquid-plastic gas-leak detectors must be used. The leak may be found by applying the solution to the suspected area and watching for bubbles to appear (see Fig. 2–75). The bubbles will appear in only a few seconds. When repairing leaks that require heating the tubing, be sure to relieve any pressure from inside the tubing being heated. This will prevent blowout, which can result in damage to the equipment and injury to the service technician.

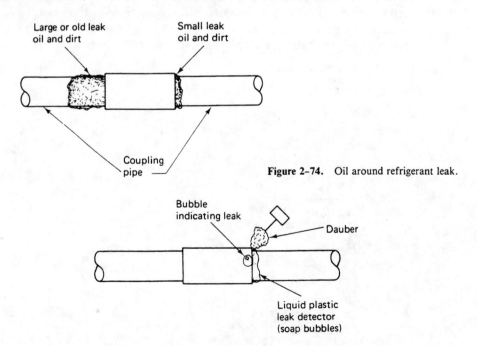

Figure 2–74.   Oil around refrigerant leak.

Figure 2–75.   Leak testing with liquid plastic leak detector.

B. Dirty Air Filters

Dirty air filters are a frequent cause of low suction pressure in an air-conditioning or heat pump system. Dirty air filters restrict the flow of air over the evaporating coil, resulting in a low load on the refrigeration system. The air filter is located on the air inlet side of the blower (see Fig. 2–76). The filters should be cleaned or changed on a regular basis to ensure peak efficiency from the unit. The throwaway-type filters should not be cleaned. They should be replaced. When cleanable-type filters are used, they should be coated with a filter coater after cleaning to increase their effectiveness.

C. Dirty Evaporating Coils

Dirty evaporating coils are often the cause of low suction pressures in air-conditioning and heat pump systems. A dirty evaporating coil restricts the airflow through the unit and the insulating value of the dirt prevents the proper transfer of heat to the refrigerant. A dirty evaporating coil is the result of a dirty air filter or one that does not fit properly. When an air filter does not fit properly, the dust-laden air will bypass it and the dust will stick to the evaporating coil fins (see Fig. 2–77). Precautions must be taken to ensure a proper fit of the air filter, and the evaporating coil must be cleaned before the system will function properly. Many times the evaporating coil will need to be removed and be steam-cleaned. Be sure to prevent moisture entering the refrigerant system during the cleaning process.

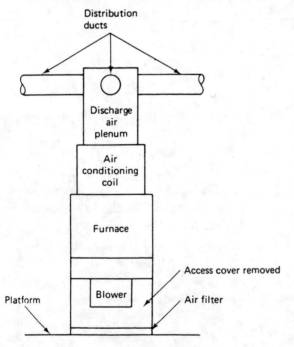

Figure 2–76. Air filter location.

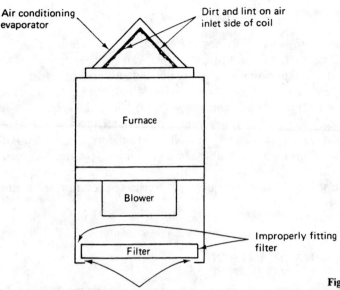

Air conditioning evaporator

Dirt and lint on air inlet side of coil

Furnace

Blower

Improperly fitting filter

Filter

Air bypass

Figure 2-77. Dirty coil because of filter air bypass.

D. Dirty Blower

A dirty blower will not deliver enough air to place the proper load on the evaporating coil, resulting in a low suction pressure. A dirty blower is the result of improperly fitting air filters or a unit that has been operated with an excessively dirty filter. The blower must be removed and the vanes cleaned so that only the metal can be seen. The filter must be cleaned, or replaced, and any air bypass must be prevented. Otherwise, the problem will reoccur. When steam-cleaning or washing the blower with water, take precautions to prevent moisture entering the electric blower motor.

E. Loose or Broken Fan Belt

A loose or broken fan belt will prevent proper operation of the blower and reduce the flow of air over the evaporating coil, resulting in a low load or no load on the evaporating coil and a low suction pressure. A loose belt that is in good condition can be properly adjusted and the unit put back in service. A broken or frayed belt or one that has become hard on the edges must be replaced and properly adjusted. To adjust a belt properly, tighten the adjustment screw until the belt can be flexed about 1 in. with one finger using moderate pressure (see Fig. 2-66). Be sure to use the proper-sized belt. A belt that is too narrow will ride the bottom of the pulley and slip, causing decreased efficiency (see Fig. 2-68). A belt that is too wide will ride too high in the pulley and will not maintain the desired efficiency and will possibly overload the motor.

F. Iced or Frosted Evaporating Coil

An iced or frosted evaporating coil will cause a low suction pressure because the insulating action of the ice or frost will prevent proper heat transfer from the air to the refrigerant. An iced evaporating coil is usually caused by insufficient load on the refrigeration system or a shortage of refrigerant. A reduced load can be the result of (1) a dirty filter, (2) a dirty evaporating coil, (3) a dirty blower, or (4) a loose or broken fan belt. A shortage of refrigerant is due to a refrigerant leak. Any of these conditions must be corrected and the coil deiced before the system will operate satisfactorily. Deicing can be accomplished by turning on the fan and turning off the compressor until the ice has been melted. A faster method is to apply a small amount of heat to the ice. Caution should be exercised to prevent overheating of the evaporating coil or the fins.

2-27 THERMOSTATIC EXPANSION VALVES

Thermostatic expansion valves are the most common type of refrigerant flow control device on heat pump units. They operate in response to both pressure and temperature inside the evaporating coil. The proper superheat adjustment of these valves will have a tremendous effect on the efficiency of the equipment.

A thermostatic expansion valve that is adjusted for too great a superheat or an improperly located feeler bulb will result in a low suction pressure. If the superheat is found to be adjusted improperly, it may be adjusted by the following procedure:

1. Measure the temperature of the suction line at the point where the feeler bulb is clamped.
2. Determine the suction pressure that exists in the suction line at the bulb location by either of the following methods:
   a. If the valve is externally equalized, a pressure gauge in the external equalizer line will indicate the correct pressure directly and accurately.
   b. Read the gauge pressure at the suction service valve of the compressor. To the reading add the estimated pressure drop through the suction line between the bulb location and the compressor service valve. The sum of the gauge reading and the estimated pressure drop will equal the approximate line pressure at the bulb.
3. Convert the pressure reading obtained in step 2a or 2b to the saturated evaporating temperature by using the pressure–temperature chart as shown in Table 2-1.
4. Subtract the two temperatures obtained in steps 1 and 3. The difference is the superheat setting of the expansion valve.

TABLE 2-1 PRESSURE–TEMPERATURE CHART[a,b]

°F	R-12	R-13	R-22	R-500	R-502	R-717 Ammonia
−100	27.0	7.5	25.6	—	23.2	27.4
−95	26.4	10.9	24.1	—	22.1	26.8
−90	25.7	14.2	23.0	—	20.7	26.1
−85	25.0	18.2	21.7	—	19.0	25.3
−80	24.1	22.2	20.2	—	17.1	24.3
−75	23.0	27.1	18.5	—	15.0	23.2
−70	21.8	32.0	16.6	—	12.6	21.9
−65	20.5	37.7	14.4	—	10.0	20.4
−60	19.0	43.5	12.0	—	7.0	18.6
−55	17.3	50.0	9.2	—	3.6	16.6
−50	15.4	57.0	6.2	—	0.0	14.3
−45	13.3	64.6	2.7	—	2.1	11.7
−40	11.0	72.7	0.5	7.9	4.3	8.7
−35	8.4	81.5	2.6	4.8	6.7	5.4
−30	5.5	91.0	4.9	1.4	9.4	1.6
−28	4.3	94.9	5.9	0.0	10.6	0.0
−26	3.0	98.9	6.9	0.7	11.7	0.8
−24	1.6	103.0	7.9	1.5	13.0	1.7
−22	0.3	107.3	9.0	2.3	14.2	2.6
−20	0.6	111.7	10.1	3.1	15.5	3.6
−18	1.3	116.2	11.3	4.0	16.9	4.6
−16	2.1	120.8	12.5	4.9	18.3	5.6
−14	2.8	125.7	13.8	5.8	19.7	6.7
−12	3.7	130.5	15.1	6.8	21.3	7.9
−10	4.5	135.4	16.5	7.8	22.8	9.0
−8	5.4	140.5	17.9	8.8	24.4	10.3
−6	6.3	145.7	19.3	9.9	26.0	11.6
−4	7.2	151.1	20.8	11.0	27.7	12.9
−2	8.2	156.5	22.4	12.1	29.5	14.3
0	9.1	162.1	24.0	13.3	31.2	15.7
2	10.2	167.9	25.6	14.5	33.1	17.2
4	11.2	173.7	27.3	15.7	35.0	18.8
6	12.3	179.8	29.1	17.0	37.0	20.4
8	13.5	185.0	30.9	18.4	39.1	22.1
10	14.6	192.1	32.8	19.8	41.1	23.8
12	15.8	198.6	34.7	21.2	43.3	25.6
14	17.1	205.2	36.7	22.7	45.5	27.5

°F	R-12	R-13	R-22	R-500	R-502	R-717 Ammonia
16	18.4	211.9	38.7	24.2	47.8	29.4
18	19.7	218.8	40.9	25.7	50.1	31.4
20	21.0	225.7	43.0	27.3	52.5	33.5
22	22.4	233.0	45.3	29.0	55.0	35.7
24	23.0	240.3	47.6	30.7	57.5	37.9
26	25.4	247.8	49.9	32.5	60.1	40.2
28	26.9	255.5	52.4	34.3	62.8	42.6
30	28.5	263.2	54.9	36.1	65.4	45.0
32	30.1	271.3	57.5	38.0	68.3	47.6
34	31.7	279.5	60.1	40.0	71.2	50.2
36	33.4	287.8	62.8	42.0	74.1	52.9
38	35.2	296.3	65.6	44.1	77.2	55.7
40	37.0	304.9	68.5	46.2	80.2	58.6
45	41.7	327.5	76.0	51.9	88.3	66.3
50	46.7	351.2	84.0	57.8	96.9	74.5
55	52.0	376.1	92.6	64.2	106.0	83.4
60	57.7	402.3	101.6	71.0	115.6	92.9
65	63.8	429.8	111.2	78.2	125.8	103.1
70	70.2	458.7	121.4	85.8	136.6	114.1
75	77.0	489.0	132.2	93.9	148.0	125.8
80	84.2	520.8	143.6	102.5	159.9	138.3
85	91.8	—	155.7	111.5	172.5	151.7
90	99.8	—	168.4	121.2	185.8	165.9
95	108.3	—	181.8	131.2	199.7	181.1
100	117.2	—	195.9	141.9	214.4	197.2
105	126.6	—	210.8	153.1	229.7	214.2
110	136.4	—	226.4	164.9	245.8	232.3
115	146.8	—	242.7	177.3	262.6	251.5
120	157.7	—	259.9	190.3	280.3	271.7
125	169.1	—	277.9	203.8	298.7	293.1
130	181.0	—	296.8	218.2	318.0	315.0
135	193.5	—	316.6	233.2	338.1	335.0
140	206.6	—	337.3	248.8	359.1	365.0
145	220.6	—	358.9	265.2	381.1	390.0
150	234.6	—	381.5	282.3	403.9	420.0
155	249.9	—	405.2	300.1	427.8	450.0
160	265.12	—	429.8	318.7	452.6	490.0

[a]Bold Figures = inches mercury vacuum; light figures = psig.
[b]Courtesy of Alco Controls Division, Emerson Electric Co.

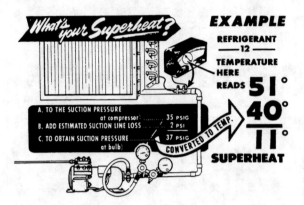

Figure 2-78. (a) Determination of superheat R-12; (b) R-22. (Courtesy of Sporlan Valve Co.)

Figure 2-78 illustrates a typical example of superheat measurement in a refrigeration system using Refrigerant-12 as the refrigerant. The temperature of the suction line at the bulb location is read at 51°F (10.6°C). The suction pressure at the compressor suction service valve is 35 psig (241.316 kPa), and the estimated pressure drop is 2 psig (117.13 kPa). The total suction pressure is 35 psig + 2 psig = 37 psig (255.106 kPa). This is equivalent to a 40°F (4.4°C) saturation temperature. Then 40°F subtracted from 51°F equals 11°F (6.2°C) superheat setting.

Notice that subtracting the difference between the temperature at the evaporating coil inlet and the temperature difference at the evaporating coil outlet is not an accurate measurement of the superheat. This method is not recommended because any evaporating coil pressure drop will result in an erroneous superheat indication.

The valve should be adjusted to provide the desired superheat setting. To determine if the expansion valve is operating, remove the remote bulb and warm it with the hand while observing the suction pressure. If the pressure increases, the valve needs adjusting.

A. Feeler Bulb on Thermostatic Expansion Valve

This is one of the major forces controlling the valve. If the bulb is located in a cold location, it will cause the valve to starve the evaporating coil, causing a lower-than-normal suction pressure. The location of the feeler bulb is extremely important and in some cases determines the succss or failure of the unit. For satisfactory expansion valve control, good thermal contact between the bulb and the suction line is essential. The bulb should be securely fastened with two bulb straps to a clean, straight section of the suction line.

Installation of the bulb to a horizontal run of the suction line is preferred. If a vertical installation cannot be avoided, the bulb should be mounted so that the capillary tubing comes from the top (see Fig. 2-79). To install, clean the suction line thoroughly before clamping the remote bulb in place. When a steel suction line is used, it is advisable to paint the line with aluminum paint to minimize future

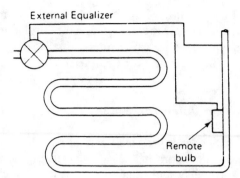

External Equalizer

Remote bulb

Figure 2-79. Remote bulb installation on vertical tubing.

corrosion and to eliminate faulty remote bulb contact with the line. On lines under ⅞ in. (22.225 mm) OD (outside diameter), the remote bulb may be installed on top of the line. On ⅞ in. OD and larger, the remote bulb should be mounted at about the 4 or 8 o'clock position (see Fig. 2-80).

If it is necessary to protect the remote bulb from the effects of the airstream after it is clamped to the line, use a material such as sponge rubber that will not absorb water when the evaporating coil temperatures are above 32°F (0°C). Below 32°F, cork or similar material sealed against moisture is suggested to prevent ice collecting at the remote bulb location.

B. Thermostatic Expansion Valve Stuck Open

When a thermostatic expansion valve is stuck open, there will be an excessive amount of sweating on the suction line. The compressor will also sweat because of the liquid refrigerant being fed back to the compressor. It is best to replace a sticking expansion valve.

C. Thermostatic Expansion Valve Too Large

If the thermostatic expansion valve is too large for the system, it will not maintain a constant suction pressure. The feeler bulb will attempt to control the flow of liquid to maintain the superheat setting, but the oversized valve will admit too much liquid too rapidly. The feeler bulb sensing this liquid will close the valve, and the pressure

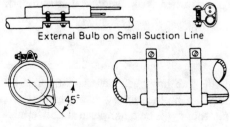

External Bulb on Small Suction Line

45°

External Bulb on Large Suction Line

Figure 2-80. Remote bulb installation on horizontal tubing. (Courtesy of Alco Controls Division, Emerson Electric Co.)

in the evaporating coil will drop until the valve opens, admitting more liquid refrigerant. This hunting action will cause the suction pressure to fluctuate, which can be seen on the compound gauge. This variation is usually 10 psig (68.9 kPa). When this condition occurs, either replace the complete valve or replace the valve seat and needle.

D. Thermostatic Expansion Valve Too Small

A thermostatic expansion valve that is too small cannot pass enough liquid refrigerant to properly refrigerate the evaporating coil. When the unit is heavily loaded, the superheat will be high and the system capacity will be low. An expansion valve that is too small will generally cause a lower-than-normal suction pressure. Either replace the valve or the valve seat and needle with the proper size.

E. Power Element of Thermostatic Expansion Valve

The power element of a thermostatic expansion valve contains a vapor charge. If a leak should develop in the power element assembly, the valve will tend to close, stopping the flow of refrigerant. To check a power element, use the following procedure:

1. Stop the compressor.
2. Remove the feeler bulb from the suction line and place it in ice water.
3. Start the compressor.
4. Remove the bulb from the ice water and warm it with the hand. At the same time feel the suction line for a drop in temperature. If the liquid refrigerant floods through the valve, the power element is operating properly. Be sure not to flood liquid refrigerant back through the suction line for a long period of time because it can cause damage to the compressor.

2-28 RESTRICTION IN A REFRIGERANT CIRCUIT

A restriction in a refrigerant circuit will reduce the flow of refrigerant, resulting in a lower-than-normal suction pressure. A restriction may be in the form of a kinked line, a plugged drier, a plugged strainer, a plugged capillary tube, or ice in the orifice of the flow control device.

A. Kinked Line

A kinked line occurs when the line has been bent too far resulting in a flattened place. This type of restriction can be located by visual inspection. However, if the restriction is in the liquid line, there will be a temperature difference across the kink.

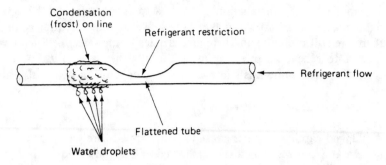

Figure 2–81.   Flattened tube causing refrigerant restriction.

If the tubing is flattened enough, there will be condensation or frost on the outside of the tubing (see Fig. 2–81).

In cases when the tube is not flattened excessively, the flattened spot may be straightened by placing a flaring block around the tube, using the proper size hole, and tightening it down. If this fails, the kinked tubing must be removed and a new piece installed. Be sure to relieve all pressure on the tubing before attempting to make repairs. Personal injury may result when repairs are attempted while there is pressure inside the tube.

B. Purpose of a Strainer

The purpose of a strainer is to trap foreign particles in the refrigeration system. A plugged strainer will reduce the flow of refrigerant and may completely stop its passage. All expansion valves are equipped with strainers. In addition, the systems have suction-line filter-driers, liquid-line driers, and a compressor suction inlet strainer. All of these features are used to trap foreign materials in the system to prevent damage to the particular component being protected. A plugged strainer will develop a pressure drop across it, and usually a temperature difference will exist on each side that can be felt with the bare hand. When a strainer becomes plugged, it is best to replace it if another one is readily available. If one is not available, thoroughly clean the original and replace it. Do not remove the strainer permanently. To do this will allow foreign particles to enter the protected device, resulting in possible damage.

C. Purpose of a Drier

The purpose of a drier is to remove moisture and trap foreign particles in the refrigeration system. A plugged drier will restrict or completely stop the flow of refrigerant through the system. All field built-up systems should be equipped with driers to remove any moisture and foreign particles that may accidentally enter the system during the installation process. A plugged drier can be determined by the temperature drop through it (see Fig. 2–82). A plugged drier is an indication that there are still moisture and foreign particles in the system. When removing a drier, be sure to relieve all pressure inside the tube being worked on. To open a refrigeration system while it is still pressurized can result in personal injury.

D. Plugged Capillary Tube

A plugged capillary tube is caused by untrapped foreign particles in a refrigeration system and can result in a reduced or completely stopped refrigerant flow through the system. A plugged capillary tube will be indicated by a longer-than-usual pressure equalization time, accompanied by a loss of refrigeration. It is recommended that a plugged capillary line be replaced rather than to clean it out. When replacing it, be sure to use the same length and inside diameter as the original. To use a different size will result in an unbalanced refrigeration system and poor refrigeration. Always replace the filter-strainer along with the capillary tube. Be sure to relieve all pressure from the system before attempting repairs or personal injury may result.

E. Ice in Orifice of Flow Control Device

This condition is due to free moisture in the refrigeration system. It usually occurs when the drier has trapped all the moisture it can absorb, and there is free moisture that turns to ice in the orifice of the flow control device. This condition is usually indicated by poor operation, a lower-than-normal suction pressure, and low discharge pressure. To be sure that moisture is causing the problem, stop the unit and apply a cloth moistened with hot water to the outlet of the flow control device. If

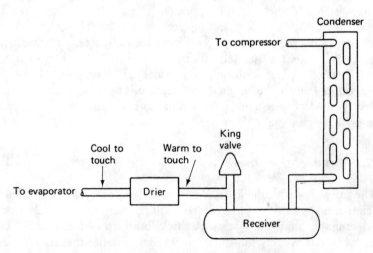

Figure 2–82.   Checking a plugged drier.

after a few minutes a hissing sound is heard and an increase in the flow of refrigerant is detected, moisture is the culprit. To correct this problem a new drier must be installed. If the problem continues, it may be necessary to purge the refrigerant from the system completely, triple-evacuate the system, install oversized driers, and recharge the system with dry refrigerant.

2-29 TUBING RATTLE

A tubing rattle is the result of mishandling or misuse of the equipment. Not only is this condition annoying, it is also damaging the tubing. If let go long enough, a refrigerant leak will appear because a hole has been rubbed in the tubing. To eliminate tubing rattle, either bend the tubing slightly apart with the hands or place a cushion between the tubes. Be careful not to apply enough pressure to the lines to break or kink them.

2-30 LOOSE MOUNTING

When mountings become worn or broken, an annoying noise will result, as well as possible damage to the unit. The only proper method of correcting this situation is to replace the mounting. In cases where a bolt or nut has become loose, it can be tightened without replacement of the mounting. When springs are used as the mountings, it is usually best to replace all the springs, not just the broken one, because the rest are probably weak and may break in the near future.

2-31 LIQUID FLASHING IN LIQUID LINE

Liquid flashing, or turning to vapor, in the liquid line may be due to a shortage of refrigerant, a liquid line that is too small, or a liquid line that has too high a vertical lift. Liquid flashing will be indicated by a hissing or gurgling noise at the expansion valve and a lower-than-normal suction pressure.

A. System Short of Refrigerant

When a system is short of refrigerant, the leak should be found and repaired and the system charged to the proper level with the proper type refrigerant and checked for proper operation. Be sure to observe all safety precautions during this process.

B. Too-Small Liquid Line

A liquid line that is too small will offer an excessive amount of resistance to the flow of refrigerant. As a result, the refrigerant pressure will drop near the outlet of the line, resulting in evaporation of a portion of the refrigerant, and picking up

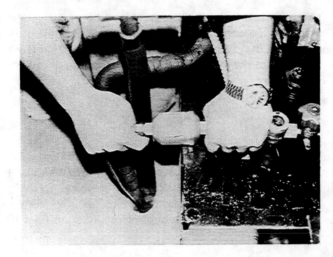

Figure 2-83. Feeling line for temperature drop.

heat outside the refrigerated space, thus reducing the effectiveness of the system. The liquid line will also drop in temperature near the evaporating coil. This drop in temperature can be felt with the hand (see Fig. 2-83). Oil return may also be a problem when this situation is experienced. When too small a liquid line is found, it must be replaced with one of the proper size. Manufacturers provide line sizing charts that can be used to size refrigerant lines properly.

C. Vertical Liquid Line

When liquid lines extend vertically for more than approximately 20 ft (6.09 m), the weight of the liquid forcing downward will probably cause flash gas. The flash gas is due to the drop in pressure. Example: If liquid Refrigerant-22 is forced up 30 ft (9.144 m), the pressure on top of the liquid will be 15 psig (206.7 kPa) lower than that at the bottom of the column. This pressure difference plus the heat in the liquid will cause flash gas. When this condition is encountered, a lower-than-normal suction pressure and a reduction in temperature of the liquid near the evaporating coil will be present. To solve this problem, the liquid refrigerant must be subcooled. This may be accomplished by installing another condensing coil in the liquid line near the outlet of the first condensing coil, or in severe cases, a suction to liquid heat exchanger should be added to the system (see Fig. 2-84). To prevent overloading the compressor, use the proper-size heat exchanger.

2-32 RESTRICTED OR UNDERSIZED REFRIGERANT LINES

Restricted or undersized refrigerant lines can be a source of continual trouble in an otherwise carefully selected unit. Often the system capacity is reduced, and oil return problems are frequently experienced with undersized lines. When these two con-

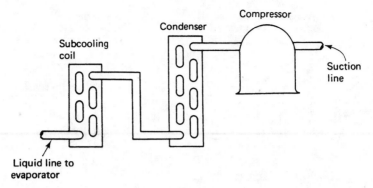

Figure 2-84. Subcooling coil location.

ditions are repeated, a reevaluation of the refrigerant line sizing method should be done. There are line sizing charts available for almost every situation encountered. These charts should be consulted and the lines replaced with the proper size, or an alternate recommended multiple set of lines installed.

A. Restricted Refrigerant Lines

Restricted refrigerant lines are usually due to contaminants or kinks in the system. A restriction in a line can be located by the difference in temperature on each side of the restriction. A restriction may be found in a drier, filter, strainer, or some other system component. This condition will be accompanied by pressures that are lower than normal and a reduction in system capacity. The restriction must be removed before proper operation can be achieved. Be sure to practice safety precautions when removing restrictions from the systems.

2-33 SUCTION PRESSURE TOO HIGH

A suction pressure that is too high is generally due to an excessive airflow over the evaporating coil, an overcharge of refrigerant, or bad suction valves in the compressor.

A. Compressor Suction Valves

The first step is to check the compressor suction valves. To do this, install a compound gauge on the compressor suction service valve (see Fig. 2-85). With the unit running, front-seat the compressor suction service valve. Observe the pressure on the compound gauge. After the pressure has fallen as low as possible, stop the compressor and observe the compound gauge. The pressure should increase only slightly. If the pressure increases to 0 psig (100.68 kPa), start the compressor again and reduce the suction pressure as much as possible. The suction pressure should be at

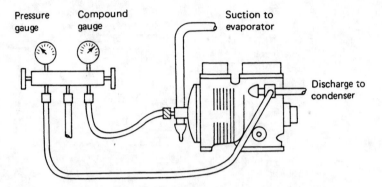

Figure 2-85.   Gauges installed on a compressor.

least a 26 in. vacuum (13.788 kPa) or lower. Stop the compressor and observe the compound gauge. The gauge should not indicate an increase in pressure of more than a 5 in. vacuum (17.25 kPa). If a greater increase than this is shown, replace the compressor valves. It is generally recommended to replace the suction valves, discharge valves, and the valve plate when replacing any one of the three. If the unit is a hermetic compressor, the complete compressor must be replaced.

B. Overcharge of Refrigerant

An overcharge of refrigerant will cause both the suction and discharge pressure to be higher than normal. An overcharge of refrigerant will cause approximately one-half of the condensing coil tubes to be cool to the touch. The cool tubes are full of subcooled liquid refrigerant, which must be removed from the system. When re-

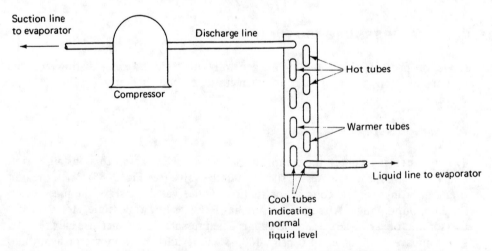

Figure 2-86.   Temperatures of a normally working condensing coil.

moving the excess, purge the refrigerant slowly and in small amounts to prevent removing too much refrigerant. When the proper charge of refrigerant is reached, only the bottom two or three rows of tubing will be cool. The others will be warm or hot to the touch (see Fig. 2-86).

C. Excessive Amount of Air Blowing Over Evaporating Coil

An excessive amount of air blowing over the evaporating coil will have the same effect as too large an evaporating coil. The system will be overloaded by the extra amount of heat absorbed into the system. To check for this condition, measure the temperature drop across the evaporating coil (see Fig. 2-87). If the temperature drop is less than that desired, the airflow must be reduced to provide the proper temperature drop. This can be done by reducing the fan speed, replacing the motor with a lower-rpm motor, or in extreme cases, blocking the return airflow will help the situation.

2-34 MOISTURE IN SYSTEM

Moisture that has entered the system can cause many serious problems. The most obvious problem is freezing of the moisture in the flow control orifice. Other more serious problems due to moisture in a refrigeration system are (1) acid forming, and (2) sludging of the compressor lubricating oil.

A. Moisture Freezing in Flow Control Orifice

This condition will result in poor refrigeration, accompanied by a lower-than-normal suction pressure and possibly a lower-than-normal discharge pressure. To determine if moisture is freezing in the flow control device, stop the compressor

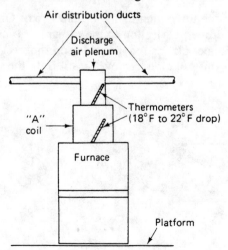

Figure 2-87. Checking temperature drop across evaporating coil.

with the compound gauge connected. Warm the flow control orifice area with a rag wet with hot water while observing the compound gauge. *Caution:* Never heat a flow control device with a welding torch, as damage to the flow control may result. If ice is causing the problem, it should melt within a few minutes, allowing the refrigerant to pass through, which will cause a rapid increase in the low-side pressure. There may also be a hissing noise in the flow control while the refrigerant is passing through. To correct this situation, install new driers in the system. It is sometimes desirable to use oversize driers to remove the moisture. In severe cases, complete discharging of the refrigerant, triple evacuation of the system, and installation of driers in both the liquid and suction lines are required, together with a complete new charge of dry refrigerant. The system should be checked at least every 24 hours for several days and the driers replaced when necessary to remove all the moisture completely.

B. Acid Forming Due to Moisture

Acid forming in a system due to moisture will cause serious damage to the compressor and expansion valve parts and will attack the insulation on the motor windings in a hermetic or semihermetic compressor. The preliminary indication of acid caused by moisture is copper plating (a copper color on the steel valves and parts) of the components (see Fig. 2–88). More advanced stages are indicated by motor burnout or repeated motor burnout. A discoloration of the oil will also be noticed. This condition may be present without freezing of the refrigerant flow control device. To correct this problem, replace the refrigerant driers and check the oil every 24 hours for several days. Replace the driers when needed until the oil returns to normal.

C. Sludging of Compressor Oil

This condition is usually due to contaminants in the system. Oil that has started sludging will not provide proper lubrication to the system. This condition will usually be accompanied by copper plating (copper color on the steel valves and parts) (see Fig. 2–88). To correct this situation, remove the oil from the system and charge

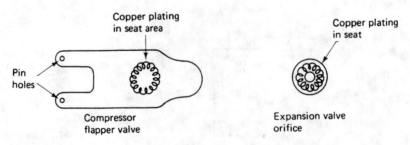

Figure 2-88.   Copper plating of steel seats.

the proper amount of fresh clean oil into the compressor. Replace all the refrigerant driers and check the system at least every 24 hours for several days. Replace the driers when needed until the oil color returns to normal.

2-35 ELECTRIC MOTORS

Electric motors are used for many purposes in heating systems other than to drive the compressor in a heat pump. The majority of uses for which these motors are used include fan motors and pump motors. There are several different types of motors in use. They are (1) split-phase (SP), (2) permanent split capacitor (PSC), (3) capacitor start (CS), (4) capacitor start, capacitor run (CSCR), and (5) shaded pole. The amount of starting and running torque required to do the job will determine the type of motor used. The steps listed below are used to check electric motors. If any of the following conditions are found, the motor must be either repaired or replaced.

A. Open Motor Windings

Open motor windings occur when the path for electric current is interrupted. This interruption occurs when the insulation on the wire becomes bad and allows the wire to overheat and burn apart (see Fig. 2-89). To check for an open winding, remove all external wiring from the motor terminals or connections. Using the ohmmeter, check the continuity from one terminal to another terminal (see Fig. 2-90).

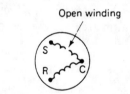

Figure 2-89.   Open motor winding.

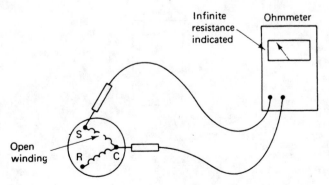

Figure 2-90. Checking continuity of open motor winding.

Be sure to zero the ohmmeter. The open winding will be indicated by an "infinity" resistance reading on the ohmmeter. There should be no continuity from any terminal to the motor case. Repair or replace the motor if found to be faulty.

B. Shorted Motor Windings

Shorted motor windings occur when the insulation on the winding becomes bad and allows a shorted condition (two wires to touch), which allows the electric current to bypass part of the winding (see Fig. 2–91). In some instances, depending on how much of the winding is bypassed, the motor may continue to operate but will draw excessive amperage. To check for a shorted winding, remove all external wiring from the motor terminals. Using the ohmmeter, check the continuity from one terminal to another terminal (see Fig. 2–92). Be sure to zero the ohmmeter. The shorted winding will be indicated by a less-than-normal resistance. In some cases it will be necessary to consult the motor manufacturer's data for the particular motor to determine the correct resistance requirements. There should be no continuity from any terminal to the motor case. Repair or replace the motor if found to be faulty.

C. Grounded Motor Windings

Grounded motor windings occur when the insulation on the winding is broken down and the winding becomes shorted to the motor housing (see Fig. 2–93). Be sure to zero the ohmmeter. The grounded winding will be indicated by a low resistance reading on the ohmmeter (see Fig. 2–94). It may be necessary to remove some paint or scale from the motor case so that an accurate reading can be obtained. If a reading is obtained, repair or replace the motor.

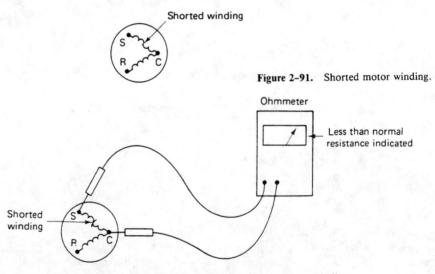

Figure 2-91. Shorted motor winding.

Figure 2-92. Checking continuity of shorted motor winding.

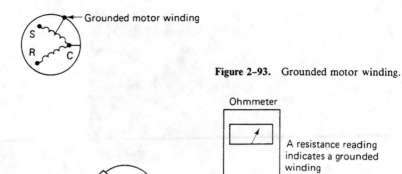

Figure 2-93. Grounded motor winding.

Figure 2-94. Checking for grounded motor winding.

D. Bad Starting Switch

A bad starting switch on a split-phase motor will prevent the proper starting of the motor. This switch provides a path for electric current to flow to the starting (aux-iliary) winding during the starting period, and it interrupts the electrical power when the motor has reached approximately 75% of its running speed. The contacts on these switches may become stuck closed, pitted, or stuck open.

A starting switch that is stuck closed will allow the motor to start but will cause the overload to open and stop the motor after a short period of operation. To check for a stuck-closed starting switch, start the motor and check the amperage draw. The amperage draw to a motor with a stuck-closed starting switch will not drop when 75% of the running speed is reached. If there is any drop in amperage, check for an overload or bad bearings. If the amperage draw does not drop, have the starting switch replaced.

A starting switch that is pitted or stuck-open will not allow the motor to start. A pitted or stuck-open starting switch will, however, allow the motor to hum and try to start. When this condition is encountered, start the motor turning while it is humming. If the motor comes up to speed, the amperage draw is normal, and the motor operates normally, have the starting switch replaced. If the motor sometimes runs in the wrong direction, the starting switch is probably the cause. Have the starting switch replaced.

E. Bad Bearings

Bad bearings will cause an overloaded condition, which will cause an excessive cur-rent draw. The motor may cut off due to the overload after operating for a while. To check for bad bearings, remove the belt if one is used, and free the motor shaft.

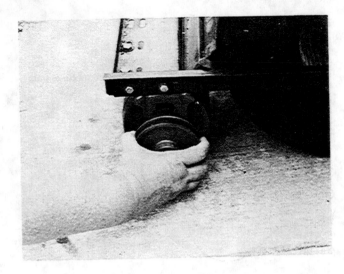

Figure 2-95. Checking motor bearings.

Move the motor shaft in a sideways movement with the hand (see Fig. 2-95). If movement in the shaft is found, either have the bearings replaced or replace the entire motor.

F. Motor Replacement

Motor replacement should be done with great care. An exact replacement must be found to be certain that efficiency is maintained. Be certain that the motor manufacturer's wiring diagram is followed. After tightening all the motor mounts, check to be sure that the shaft is free to rotate. Start the motor momentarily to see that it turns in the right direction and that nothing is dragging. Then start the motor and allow it to operate under normal conditions while checking the amperage draw. If the amperage draw is excessive, the reason should be found and corrected before leaving the job.

2-36 TRANSFORMERS

Transformers are electrical devices that are used to reduce, or increase, electrical voltage. In heating work, they are used to reduce line voltage to low voltage (24 V) (see Fig. 2-96). The 24 V is used in the control circuit because it is safer, the controls are cheaper to manufacture, and the controls are more responsive than line voltage controls to temperature change. To check a transformer, use the voltmeter and check the input voltage on the primary side of the transformer (see Fig. 2-97). If voltage is found, check the output voltage from the secondary side of the transformer (see Fig. 2-98). If no secondary voltage is found and primary voltage is found, the transformer is bad and must be replaced. If no voltage is found to the primary side of the transformer, the trouble is elsewhere. An alternate method of checking a trans-

Figure 2-96. Class 2 transformer.

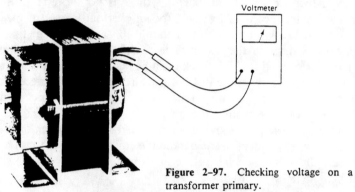

Figure 2-97. Checking voltage on a transformer primary.

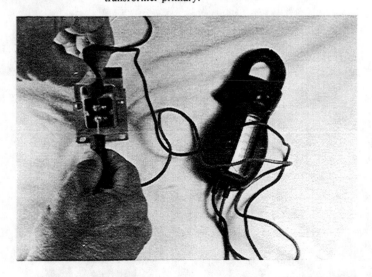

Figure 2-98. Checking voltage on a transformer secondary.

former is first to disconnect all external wiring to the transformer. Then check the primary and secondary windings, in turn, with an ohmmeter. If either winding is open, the transformer is bad and must be replaced.

Some transformers incorporate a fuse in the secondary winding. If this fuse should blow, the transformer is rendered inoperative and usually must be replaced. Fuses can be replaced in only a few transformers.

A. Replacing a Transformer

When replacing a transformer, be sure to use one with at least an equal VA (volt-ampere) rating as the one being replaced. Always use a replacement designed for the same electrical characteristics as the one being replaced. Never use one with a smaller VA rating unless it is known that the replacement will have sufficient capacity. A transformer with too small a VA rating will only burn out because of an overload condition. Always check to be sure that there is no short to cause the replacement transformer to burn out. To check for an overload or short, measure the amperage draw in the low-voltage circuit. Then multiply the amperage times the voltage to obtain the VA draw of the circuit. The circuit VA must be equal to or smaller than the transformer rating. This check must be made with the unit running. Because of the small current draw in the low-voltage circuit, it may be necessary to use a multiplier. These multipliers may be purchased or may be handmade. To make one, simply coil a piece of wire around the tong of an ammeter (see Fig. 2-99). Connect the multiplier into the circuit and check the current draw. Then divide the current draw by the number of turns in the handmade coil. *Example:* A coil has 10 turns and the current draw is 4.5 A. Therefore, 4.5 divided by 10 = 0.45 A. The VA rating would be 0.45 × 24 = 10.8 VA.

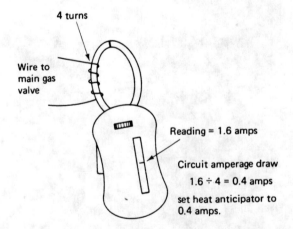

Figure 2-99.   Checking amperage in a temperature control circuit.

2-37 *HIGH AMBIENT TEMPERATURES*

High ambient temperatures are usually encountered in the summertime, especially in air-conditioning installations. High ambient temperatures result in high discharge pressure and poor operation of the unit. The best method of helping this situation is to provide a shade of some type over the unit. The shade should extend past the unit toward the direction of airflow (see Fig. 2-100).

An alternate method is to provide a spray of water on the condensing coil. This can involve an intricately designed unit or a lawn sprinkler, depending on the efficiency required (see Fig. 2-101). Care should be taken to protect the electric motors and components from becoming electrically shorted to the unit.

2-38 *HIGH RETURN AIR TEMPERATURE*

High return air temperature is usually encountered when the unit has been shut down for a while and then is restarted. The high temperature of the air flowing over the evaporating coil will cause a high suction pressure, resulting in temporary overload of the unit. The only solution is to allow the unit to operate until the return air temperature is lowered to the designed operating temperature. The unit should then operate normally.

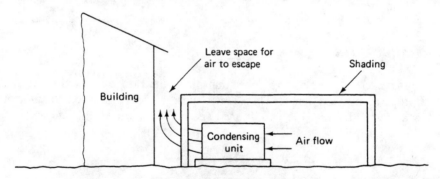

Figure 2-100.   Shading for outdoor unit.

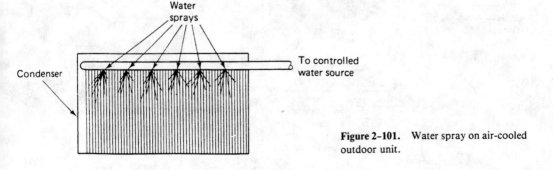

Figure 2-101.   Water spray on air-cooled outdoor unit.

2-39 INDOOR FAN RELAY

The purpose of an indoor fan relay is to allow the user to select continuous or intermittent operation of the indoor fan motor at the thermostat, and allow the fan to operate on a slow speed during the heating season and on high speed during the cooling season. This type of relay is normally equipped with one set of normally open and one set of normally closed contacts. There are two general types of fan relays (see Fig. 2-102).

The shrouded type has marked terminals and the open type has connections directly onto the contact blades. To check out a fan relay, first turn the thermostat fan switch to "on." The relay should click. If there is no click, check the voltage on the two coil connections (see Fig. 2-103). If voltage is indicated and no click is heard when the relay is energized, the relay is sticking and should be replaced. If a click is heard and the fan still does not start, check the line voltage across the common connection and the other two connections in turn (see Fig. 2-104).

With the relay deenergized, a voltage reading should be obtained between the common and low-speed connections. If these two checks are not proven, the relay is bad and should be replaced. However, if these two checks prove the relay is good, the problem is elsewhere and should be found. When the relay must be replaced, be sure that the replacement has the same voltage and amperage ratings as the one being replaced.

2-40 AIR DUCTS

Air ducts are the hollow tubes that direct the flow of air from the air handler to the conditioned space. These ducts must be properly sized to direct the desired amount of air to the conditioned space. To design a practical and efficient duct

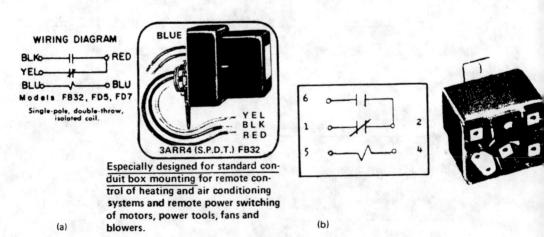

Figure 2-102.   Shrouded (a) and open-type (b) fan relays.

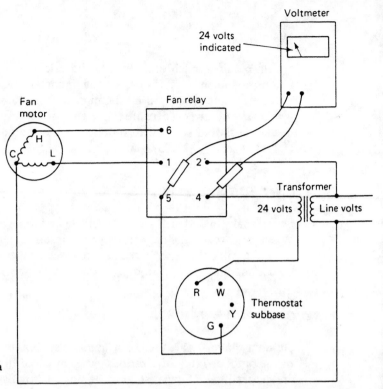

Voltmeter

24 volts
indicated

Fan
motor

Fan relay

Transformer

24 volts   Line volts

Thermostat
subbase

Figure 2–103. Checking voltage on a
fan relay coil.

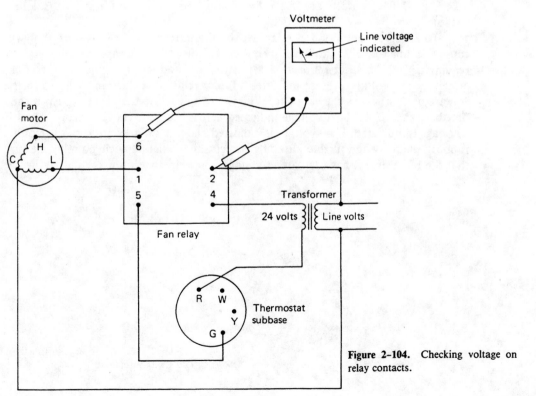

Voltmeter

Line voltage
indicated

Fan
motor

Transformer

24 volts   Line volts

Fan relay

Thermostat
subbase

Figure 2–104. Checking voltage on
relay contacts.

system requires much time, effort, and calculations. The proper amount of insulation around the ducts is one of the important features that requires careful calculation. There should be a minimum of 2 in. (25.4 mm) thickness of insulation with a vapor barrier. There are times when more insulation may be required. Enough insulation should be applied to prevent a heat loss or gain of more than about 2°F (1.11°C) (see Fig. 2–105).

2–41 UNIT TOO SMALL

Occasionally, a designer will miscalculate and the wrong-sized unit will be installed. When this occurs, the only remedy is to install a unit of proper size. A unit that is too small will operate properly, but the space temperature will be higher than desired on cooling and lower than desired on heating (see Sec. 2–50).

2–42 OUTDOOR FAN RELAY

Outdoor fan relays are used on heat pump systems to aid in the starting and stopping of the outdoor fan in the various cycles involved in this type of system. The relays are generally SPST-type (single-pole single-throw) switches, which are actuated by a 24-V coil. The contacts are in the line voltage circuit to the fan motor (see Fig. 2–106).

To check an outdoor fan relay, set the thermostat to a temperature that will cause the unit to run. With a voltmeter, check the voltage across the relay coil. A reading of 24 V should be indicated. Cycle the unit and listen for the relay to click. No click with voltage present indicates a faulty relay. If a click and voltage to the coil are present, check the contacts. With the relay energized and the system turned on, there should not be a voltage indicated across the contacts. The relay is bad if a voltage is indicated. If no voltage is indicated and the fan does not run, the trouble is not in the relay and further checking is necessary. When replacing a relay, be sure that the replacement has the correct voltage and amperage rating.

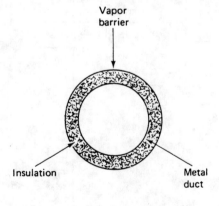

Vapor
barrier

Insulation

Metal
duct

Figure 2–105.   Cross section of metal duct with insulation.

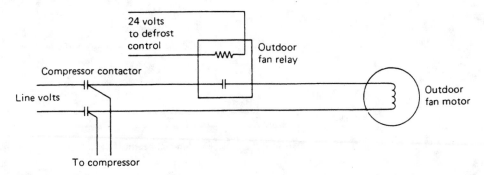

Figure 2-106. Outdoor fan relay connections.

2-43 *HEATING RELAY*

Heating relays are used on electric furnaces to complete the electrical circuit to the heating elements on demand from the thermostat. These relays operate by an electrically operated bimetal warp switch (see Fig. 2-107). The bimetal heater is in the 24-V circuit, and the contacts are in the line voltage to the heating elements (see Fig. 2-108).

To check the heating relay, use a voltmeter to check the voltage across the contacts. If a voltage is indicated here, the contacts are open. Next, jumper the contacts. If the heating element is energized, the trouble is in the relay. An energized heating element will be indicated by a current flow to the element. Replace the relay with an exact replacement or the unit will not operate as designed.

Figure 2-107. Bimetal heating relays.

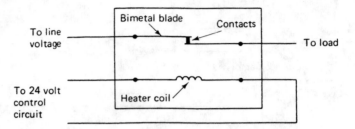

Figure 2-108. Inside of bimetal heating relay.

2-44 FAN CONTROL

The fan control is used on heating systems to start and stop the fan motor on an increase or decrease in temperature inside the furnace. This control is manufactured in adjustable and nonadjustable types as well as in extended element and bimetal disk types. Some manufacturers use electric time-delay relays for fan controls on some units. The bimetal types are set to bring the fan on when the sensing element temperature reaches 135°F (57°C) to 150°F (66°C), and stop the fan when the temperature drops to approximately 100°F (37.8°C). The electric types are preset for a specific time to lapse for the "on" and "off" settings. To check out a fan control, insert a thermometer in the unit as close as possible to the sensing element (see Fig. 2-109).

Start the main burner and observe the thermometer. When the desired temperature is reached or the specified time has lapsed, the fan should start. If not, adjust the control if it is of the adjustable type. Next, stop the main burner and observe the thermometer. When approximately 100°F (37.8°C) is reached, the fan should stop. If it does not stop, adjust the control if it is of the adjustable type. If

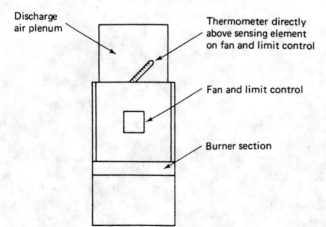

Figure 2-109. Checking temperature setting of a fan control.

the control is not adjustable, or more than 15°F (8.33°C) of adjustment is needed, replace the control. Be sure to use an exact replacement or the unit will not operate as designed. Never attempt to make internal adjustments on these controls because of the danger of overheating and permanently damaging the equipment, possible fire, and personal injury (see Section 4–4).

A. Fan Cycles When Main Burner Is Off

If the fan should cycle when the main burner is off, check to see that the airflow is not restricted by a dirty filter or dirty evaporating coil. When no air restriction is found, widening the differential between the fan "on" and "off" settings will usually solve the problem.

B. Fan Cycles When Main Burner Stays On

If the fan cycles on and off when the main burner remains on, the air temperature may be too low or too much air may be blown through the furnace. Allow the room temperature to increase and reduce the airflow to provide the desired temperature rise, as indicated on the furnace nameplate. Sometimes, widening the differential between the fan "on" and "off" settings will solve the problem.

2-45 HEAT SEQUENCING RELAYS

Heat sequencing relays are used on electric furnaces to prevent all of the heating elements coming on or off at the same time. Therefore, the relays may be used to vary the capacity of the unit. There are three basic types of relays used to accomplish sequencing: (1) thermal element switches, also known as bimetal time-delay switches, which heat up and close circuits; (2) relays that activate another relay along with the heating element; and (3) modulating motors that rotate to close contacts at various points in the degrees of rotation.

A. Thermal Element Switches

Thermal element switches make use of a bimetal heated by an electric current to actuate the contacts (see Fig. 2–107). These switches are generally termed stack relays. They have one heated bimetal which closes more than one set of contacts. The use of the different contacts will vary depending on the equipment design. One set of contacts may complete the electrical circuit to the fan motor; another set may complete the electrical circuit to the heating element; and yet another set may complete the electrical circuit to the heating element of another relay. The specific use will need to be determined on each unit.

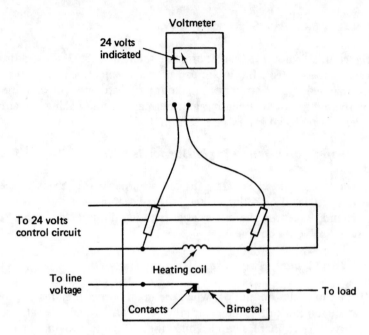

Figure 2-110. Checking voltage across thermal element on a time-delay relay.

To check these switches, first set the thermostat to demand heating. Then listen to the switch. If a click is heard, the heating element is working. If no click is heard, check the voltage across the heating coil (see Fig. 2-110). If a voltage is indicated, the heating element is bad and the switch must be replaced. If a click is heard, indicating that the heating element is working, check the voltage across the contacts. If a voltage is indicated, the contacts are bad and the switch must be replaced. Be sure that full electrical power is supplied to the unit. The replacement relay must be an exact replacement or the unit will not function as designed.

B. Time-Delay Relays

Time-delay relays that actuate relays along with the heating element are also some-times used. These are single-contact relays. When these relays are used, the relays energized by these devices must be equipped with heating elements of the same volt-age as the heating element energized by the original relays. The checkout procedure is the same as that described for thermal element switches.

C. Modulating Motors

Modulating motors are generally used on more sophisticated units requiring more exact control than other units. These motors will operate in any position within a number of degrees of rotation for which they were designed. The motor is controlled by use of a Wheatstone bridge (see Fig. 2-111). The motor shaft is extended through the housing. The switches are placed on the shaft to complete the desired circuit at a given degree of rotation. The thermostat controls the motor by use of potentiometers, which produce a balanced voltage effect and control the balancing relay in the motor. To check out these motors, first be sure that the required voltage is being supplied to the motor and controls. Next, set the thermostat to demand heating. The motor should start turning. If not, clean the potentiometers and the balancing relay contacts with a cleaning solvent. It is not generally recommended to burnish a potentiometer because of possible damage. If the motor still does not operate, check to see that voltage is supplied to the motor. The manufacturer's wiring diagram may be required to determine the motor terminals. If voltage is indicated at these terminals and the motor still does not operate, replace the motor. If the motor operates but the heating elements do not function, check the voltage across each switch after making certain that the correct voltage is available. If voltage is indicated across the contacts, the switch is defective and must be replaced. Be sure to install the replacement switch in the exact position as the replaced switch, to maintain equipment efficiency.

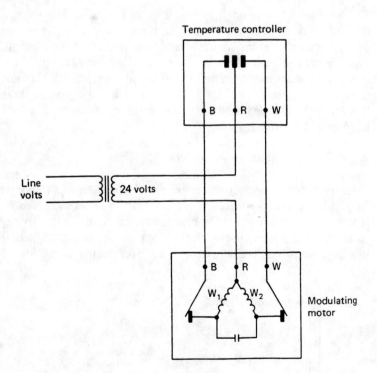

Figure 2-111.   Wheatstone bridge connections for modulating motor.

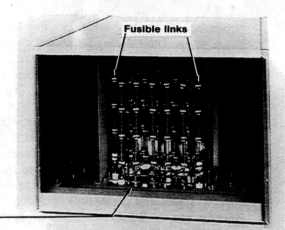

Heating elements are staggered for more uniform heat transfer, eliminates hot spots, and
insures black heat. Each element is protected
by an over temperature disc in one end and a
fuse link in the other.

Figure 2–112.   Open-wire-type heating element. (Courtesy of INDEECO.)

2–46 HEATING ELEMENTS

There are several types of heating elements available. However, the open wire element is the most popular. Heating elements generally are equipped with protective
devices in case of excessive current or overheating of the element (see Fig. 2–112).
These heating elements produce heat when electricity is applied to them. Sometimes
they will develop hot spots and the wire will separate, causing an open circuit, and
the heating will stop (see Fig. 2–113). A visual inspection of the element will show
this condition. A separated heating element must be replaced before normal operation can be resumed.

A. Over-Temperature Protective Devices

Over-temperature protective devices are required by UL (Underwriters' Laborato-
ries) and the NEC (*National Electrical Code*®). There are two types of protection
required by both agencies. The primary system is automatically reset and is designed
to interrupt the flow of electrical current to the heater if insufficient airflow, air
blockage, or other causes result in an overheating condition. The secondary or

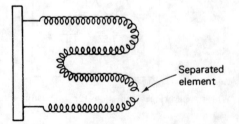

Separated
element

Figure 2–113.   Separated electric heating
element.

Figure 2-114.   Element protective device. (Courtesy of INDEECO.)

backup system is designed to operate at a higher temperature and interrupt the flow of electrical current to the heater if the primary system should fail or continued operation under unsafe conditions.

When standard slip-in heaters are used, either a disk type or a bulb and capillary thermal cutout may be used for both the primary and secondary system of temperature protection (see Fig. 2-114). While this type of cutout is standard in the field for primary protection, some manufacturers also use them for control of the secondary system. The fusible links require field replacement, but the manual reset controls can immediately be put back in operation simply by pressing the reset button located on the heater terminal box. Be sure to determine and correct the cause of the trouble before leaving the job.

B. Overcurrent Protection

Overcurrent protection is required on all electrical devices by UL and NEC. These agencies require that an electric heater drawing more than 48 A total must be subdivided into circuits that draw no more than 48 A each. Each circuit must be protected by fuses or circuit breakers. Because electric heaters are regarded as a continuous load, overcurrent protective devices must be rated for at least 125% of the circuit load. Thus, while the circuits are limited to 48 A, the fuses are limited to 60 A. These fuses must be applied by the heater manufacturer. When a fuse is blown there is an electrical overload, which must be found and removed before the service job is complete.

C. Loss of Airflow Protection

Loss of airflow protection in an electric heater is required by both UL and NEC. Both agencies require that each heater be electrically interlocked with the fan so that the heater cannot operate unless the fan circuit is energized and air is therefore flowing (see Fig. 2-115). Most slip-in heaters include a built-in differential pressure, diaphragm-operated, airflow switch to meet this requirement. A loss of airflow may be caused by dirty filters, dirty blower, an inoperative blower, or air outlets closed off. The problem must be corrected before the service job is complete.

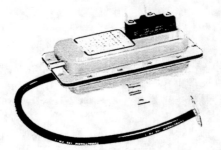

Figure 2-115. Airflow protective device. (Courtesy of INDEECO.)

2-47 TEMPERATURE RISE THROUGH A FURNACE

The amount of temperature rise through a furnace is recommended by the furnace manufacturer. The minimum and maximum limits are designated on the furnace nameplate. When these limits are exceeded, the unit will not perform as it was designed. The temperature rise may be determined by subtracting the inlet air dry bulb temperature from the outlet air dry bulb temperature. To obtain these temperatures, insert a thermometer as close to the air inlet and outlet of the furnace as possible (see Fig. 2-116). When the maximum temperature rise is reached, the minimum of air is flowing through the furnace. If this limit is exceeded, damage to the heater

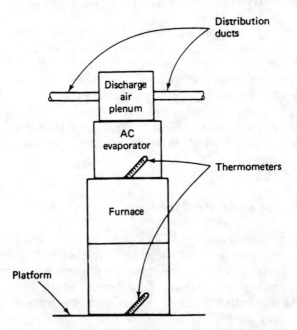

Figure 2-116. Checking temperature rise through a furnace.

may occur. When the minimum temperature rise is reached, the maximum amount of air is flowing through the furnace. If this limit is exceeded, cold drafts may be experienced. When operating on a maximum temperature rise, the relative humidity will be lowered. When operating on a minimum temperature rise, the relative humidity will be raised. To change the temperature rise, adjust the fan speed to deliver the amount of air required.

2-48 BLOWERS

Blowers are the devices used to force air through the equipment and into the conditioned space. Blowers are generally located at the air inlet to the furnace or to the air distribution system. A blower will not move a sufficient amount of air if the vanes become clogged with dirt, if the belt becomes loose and slipping, if the bearings become worn or out of lubricant, or if the motor develops a problem.

A. The Blower

The blower must be removed from the equipment for it to be cleaned effectively. Sometimes a brush and soapy water are required to clean a blower. In such cases the blower should be taken out of the building and the motor protected from the water spray.

B. Defective Bearings

Defective bearings will overload the motor and will cause the blower to turn more slowly than is required. Therefore, the airflow will be reduced, affecting equipment operation. To check for defective bearings, remove the fan belt and grasp the blower shaft in the hand. Move the shaft from side to side (see Fig. 2-117). If sideways

Figure 2-117.   Checking fan bearings.

movement is detected, replace the bearings. Be sure to check the blower shaft for wear or scoring. If wear or scoring is indicated, replace the shaft also. If no sideways movement is detected, spin the blower and observe if the blower turns free or if it is stiff. If a stiffness is indicated, lubricate the bearings. If lubrication does not solve the problem, replace the bearings as described above. Do not overtighten the belt, or excessive bearing wear will result. Belt adjustment is discussed in Section 2–19. Motor service is discussed in Section 2–36.

2-49 BTU INPUT

The Btu input to an electric heating unit can be changed in the field without much problem. When more or less heat is required, change the number of elements that are in operation. If more heat is required, add enough heating elements to supply the demand. When less heat is required, remove the extra heating elements. The Btu input to a gas furnace cannot be increased above the Btu input rating of the furnace. To do so would cause overheating of the heat exchanger and would also upset the vent system. The Btu input to a gas furnace can be reduced by only 20% of the input rating. To reduce the input Btu more would upset the vent system. An improperly operating vent system is very dangerous and must be avoided. When changing the Btu input of any furnace, be sure to check the temperature rise through the furnace as described in Section 2–47. Make any adjustments on the blower speed to maintain the recommended temperature rise. In some cases a larger blower will be required.

2-50 HUMIDITY

During the winter months a desired relative humidity inside a conditioned space will be between 40 and 60%. However, when the outdoor temperature is very low, condensation on doors, walls, and windows may be experienced with this humidity level. This is an indication of excessive relative humidity, which must be lowered to prevent damage. This humidity may be the result of cooking with open pots and pans or bathing. When cooking with open pots and pans, the pans should be covered to prevent the escape of moisture into the conditioned space. If covering the pots is not desirable, a vent system may be installed above the stove. When bathing causes the problem, install a vent system in the bathroom. Thus vents can be used when the cooking or bathing is in progress.

2-51 PILOT SAFETY CIRCUIT

The pilot safety circuit is the device that makes fuel burning equipment safe. The components that make up the pilot safety circuit are the thermocouple, the pilot burner, and the pilot safety device. These devices are designed to stop the flow of gas to the main burners when the pilot flame is not sufficient to ignite the main burner gas safely.

A. Thermocouples

Thermocouples are devices that when heated produce a small electrical current (see Fig. 2–118). This small current (approximately 30 mV) is used to energize an electromagnetic coil in the pilot safety device. The heat is provided by the pilot burner flame. The cartridge end of the thermocouple is mounted in the pilot burner so that the cartridge is in the pilot flame from ⅜ to ½ inch (9.524 to 12.7 mm) (see Fig. 2–119). When the cartridge is heated to a dull red, almost black color, the maximum output is gained from the thermocouple. There are three checks that can be made on a thermocouple: (1) the open-circuit test, (2) the closed-circuit test, and (3) the dropout test. All three tests are made with a dc millivolt meter (see Section 4–3).

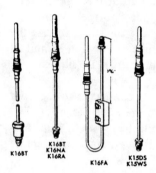

K16BT   K16BT K16NA K16RA   K16FA   K15DS K15WS

Figure 2–118. Thermocouples. (Courtesy of Johnson Controls, Inc., Control Products Division.)

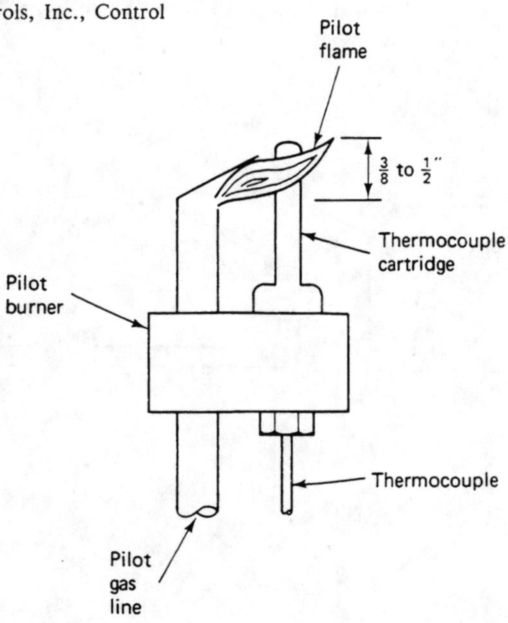

Pilot flame

$\frac{3}{8}$ to $\frac{1}{2}''$

Thermocouple cartridge

Pilot burner

Thermocouple

Pilot gas line

Figure 2–119. Thermocouple properly mounted on pilot burner.

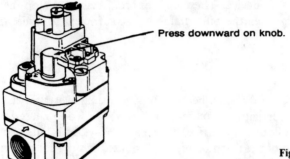

Press downward on knob.

Figure 2-120. Pressing knob on gas valve to light pilot.

The open-circuit test is made while the thermocouple is being heated with the pilot flame. The first step is to keep the pilot flame burning by depressing the bypass knob on the pilot safety device (see Fig. 2-120). Then disconnect the thermocouple from the pilot safety device and check the thermocouple output (see Fig. 2-121). Be sure to connect the meter leads properly or a reading may not be obtained. If the meter does not indicate a voltage or the needle moves off the scale, reverse the meter leads. The voltage reading obtained in this test should be between 25 and 35 mV. If the reading is as low as 20 mV, replace the thermocouple. Be sure that all connections are clean.

The closed-circuit test is made with the thermocouple connected to the pilot safety device with an adapter while the cartridge end is in the pilot flame (see Fig. 2-122). This test is used to check the thermocouple while being used. The thermocouple output should be 17 mV or more. If the output is less than 17 mV, replace the thermocouple.

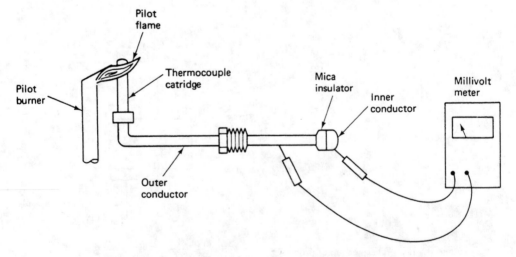

Figure 2-121. Open-circuit thermocouple test.

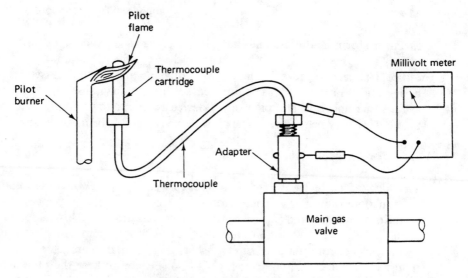

Figure 2-122. Closed-circuit thermocouple test.

The dropout test is made to determine at what voltage the pilot safety device will close when the pilot flame is out. The dropout test is made with the thermocouple connected to the pilot safety device with an adapter while the cartridge end is in the pilot flame. A millivolt meter is connected to the adapter (see Fig. 2–122). Next turn off the pilot and observe the meter reading. When the pilot safety device drops out or closes, there will be a thud in the valve and the meter needle will jump. The recommended dropout voltage is 8 mV. However, most pilot safety devices will not drop out until the voltage drops to approximately 4 mV. If the dropout voltage is as low as 2 mV, replace the pilot safety device. A pilot safety device that will not drop out above 2 mV is unsafe because it will allow excessive raw gas to escape.

B. Pilot Burner

A pilot burner has two functions: (1) to provide heat for the thermocouple, and (2) to ignite the main burner gas (see Fig. 2–123). The only thing that can happen to a pilot burner is the orifice may become clogged. When this happens, the pilot

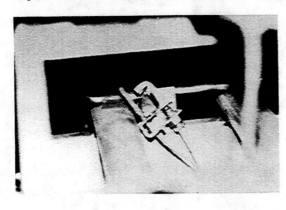

Figure 2-123. Pilot burner installation.

cannot perform its duties. The flame will become small, yellow, or both. The only remedy is to remove the pilot burner from the equipment, dismantle, and clean the orifice. Use care to prevent enlarging or damaging the orifice. Reassemble the pilot burner and reinstall it on the equipment. When a pilot flame repeatedly goes out and the thermocouple and pilot safety device are found to be good, shielding of the pilot burner may be required. The pilot flame is possibly being blown out by drafts.

C. Pilot Safety Device

The pilot safety device is used to prevent raw gas escaping when the pilot flame is not functioning properly. There are two checks to make on a pilot safety device: (1) the dropout test, and (2) the continuity test. The dropout test was outlined earlier in this section. The continuity test is made with an ohmmeter. First remove the thermocouple from the pilot safety device; then with an ohmmeter check the continuity of the electrical coil (see Fig. 2–124). Be sure to zero the ohmmeter. If no continuity is indicated or the dropout test fails, replace the pilot safety device.

2–52 ORIFICES

The purpose of an orifice is to regulate the flow of gas to the main burner. The orifice is sized according to the type of gas, the manifold pressure, and the Btu rating of the burner and gas (see Fig. 2–125). Usually, orifices are trouble-free and need little attention. However, if someone tried cleaning the orifice and ruined it, the orifice must be replaced. When replacing an orifice be sure to use an exact replacement. Often the furnace and the type of gas used are different. When this happens, the orifice must be sized and replaced for the type of gas used. There are tables for every type of gas, gas pressure, and Btu rating thinkable (see Table 2–2).

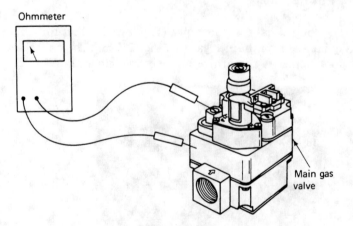

Ohmmeter

Main gas
valve

Figure 2–124. Checking continuity of pilot safety coil.

Figure 2-125. Burner orifices.

TABLE 2-2 ORIFICE CAPACITIES FOR NATURAL GAS[a]

Wire Gauge Drill Size	Rate	
	ft³/hr	Btu/hr
70	1.34	2,340
68	1.65	1,650
66	1.80	1,870
64	2.22	2,250
62	2.45	2,540
60	2.75	2,750
58	3.50	3,050
56	3.69	3,695
54	5.13	5,125
52	6.92	6,925
50	8.35	8,350
48	9.87	9,875
46	11.25	11,250
44	12.62	12,625
42	15.00	15,000
40	16.55	16,550
38	17.70	17,700
36	19.50	19,500
34	21.05	21,050
32	23.70	23,075
30	28.50	28,500
28	34.12	34,125
26	37.25	37,250
24	38.75	39,750
22	42.50	42,500
20	44.75	44,750

[a]100 Btu/ft³, manifold pressure $3\frac{1}{2}$-in. water column.

The local supplier will usually have this information and can drill the orifice to the proper size. Drill the orifice from the rear so as to center the orifice in the spud. Use care not to elongate the orifice and ruin the spud.

A. Orifices That Are Too Large

Orifices that are too large will admit excess gas to the burner. This excess gas will not burn properly and will cause carbon monoxide, a deadly gas, to be produced. A large orifice will cause a larger than desired flame, which will float off the main burner (see Fig. 2-126). Sometimes the flame will be all blue and sometimes it will be partially yellow.

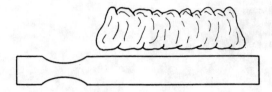

Figure 2-126. Floating main burner flame.

B. Misaligned Orifice

Misaligned orifices will misdirect the gas down the burner. The gas stream will often strike the burner, slowing down the velocity (see Fig. 2–127). This reduction in velocity will reduce the amount of primary air drawn into the burner and will prevent proper mixing of the gas and air. This condition is accompanied by a yellow flame. The orifice must be replaced or the manifold redrilled to correct the problem. Dirty orifices will often misdirect the gas stream. The remedy here is to clean the orifice. *Caution:* Be careful not to damage the orifice when cleaning.

C. Orifices That Are Too Small

Orifices that are too small will cause a smaller-than-normal flame (see Fig. 2–128). The flame will be all blue but will not provide the required amount of heat. Orifices that are too small may be redrilled to the proper size. When drilling an orifice, insert the drill from the rear of the orifice and use extreme care to drill a round hole (see Fig. 2–129). Use an orifice chart to determine the correct size.

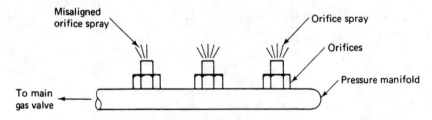

Figure 2-127. Misaligned orifice gas stream.

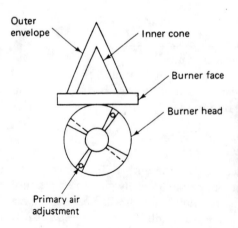

Figure 2-128. Small flame.

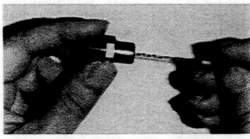

Figure 2-129. Drilling an orifice.

2-53 *MAIN BURNERS*

The purpose of a main burner is to provide a place for the mixing of gas and air and a place for the burning of the mixture (see Fig. 2–130). There are several conditions that can affect the operation of a burner, such as distorted carryover wing slots, dirty burner, primary air adjustments, poor flame travel to the burner, and poor flame distribution.

A. Carryover Wing

The carryover wing on a main burner provides a path for the pilot flame to travel from burner to burner, providing smooth gas ignition (see Fig. 2–131). If the slots in the carryover wing should become distorted, the ignition flame will not travel properly and flashback (gas burning at the orifice) will occur. The solution to this problem is to repair the slots. This is done by restoring them to their original condition. When repairing is not practical, the burners should be replaced.

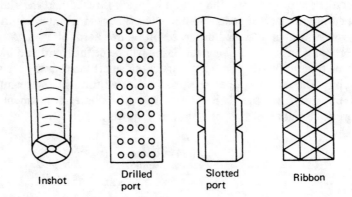

Inshot   Drilled   Slotted   Ribbon
                port   port

Figure 2-130.   Types of main burners.

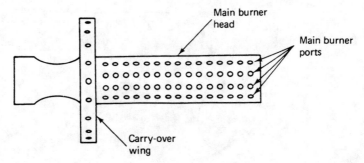

Figure 2-131.   Main burner carryover wing.

B. Gas Burners

Gas burners become dirty through use. This is because the air required for combustion contains dust and lint (see Fig. 2–132). Dust and lint particles from the air, and rust particles from the heat exchanger, collect in the burner and upset the flow of the gas and air mixture, which upsets the normal combustion process. Dirty burners are the cause of flashback (gas burning at the orifice) and possible delayed ignition. To correct this problem, remove the burners and clean both inside and outside. The burner ports must be cleaned with extreme care to prevent damage. Air pressure may be used to blow out the inside of the burner to remove any collected dirt and lint. The burners should be removed from the furnace to aid in cleaning.

C. Primary Air Adjustment

The primary air adjustment (shutter) is located on the face of the burner and is used to regulate the amount of primary air being drawn into the burner (see Fig. 2–133). The amount of air being drawn into the burner determines the type of flame present. When the shutter is off, the primary air is restricted and the flame will be yellow, indicating poor combustion. When the primary air shutter is fully open, too much air will be drawn into the burner, resulting in a hard, blowing flame, also indicating improper combustion. The ideal flame will be neither hard and blowing nor contain yellow tinges. To adjust a main burner, close off the primary air shutter and obtain a yellow flame. Next, open the primary air adjustment shutter until the yellow has disappeared from the flame. Then open the shutter adjustment about one-eighth more (one-eighth of the total shutter adjustment).

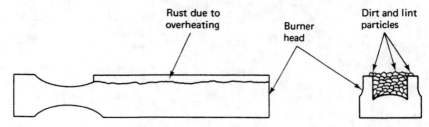

Figure 2–132.   Dirty burner.

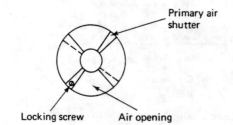

Figure 2–133.   Primary air adjustment.

2-54 GAS VALVES

Gas valves are used to control the flow of gas to the main burner. Usually, there are four main parts, each part having its own function: (1) the manual gas cock, (2) the pilot safety device, (3) the pressure regulator, and (4) the main operator (see Fig. 2–134). The manual gas cock is used to shut down the pilot and main burner manually. The pressure regulator is used to supply gas to the main burner at a constant pressure. The main operator controls the flow of gas to the main burners in response to the thermostat demands. The main operator part of these valves is in the 24-V control circuit. The pilot safety part is connected to the thermocouple and is used to shut down the pilot and main burner when the pilot flame is too small to light the main burner gas safely. These valves are usually very trouble-free and should be carefully checked before replacement. To check the valve, use a voltmeter and measure the voltage across the coil (see Fig. 2–135).

Be sure that there is power on the secondary side of the transformer and that the thermostat is demanding heat. If 24 V is present and the pilot safety circuit is functioning, the gas valve should be replaced (see Section 2–51). Also, if the gas

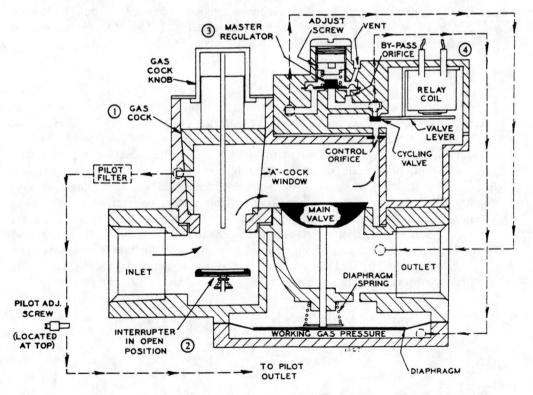

Figure 2–134.   Main gas valve cutaway. (Courtesy of White-Rodgers Division, Emerson Electric Co.)

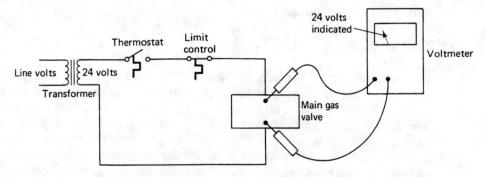

Figure 2–135. Checking voltage on a main gas valve coil.

valve operates properly at times and not at other times, the valve should be replaced. If a valve of the same physical size is used as the replacement, no plumbing changes will be needed. Do not attempt to repair gas valves. Either replace components or the complete valve.

A. Quick-Opening Gas Valves

Quick-opening main gas valves cause rough ignition and flame roll out. To prevent this, install a surge arrestor in the vent line (see Fig. 2–136). Do not attempt to repair any gas valve. Either replace components or the complete valve.

B. Redundant Gas Valves

The purposes of these valves are to regulate the flow of gas to the main burner and to provide safety. They contain two independently operated main gas valves. If one valve should fail to close, the other will close and shut off the gas flow to the main burner (see Fig. 2–137). This type of valve is used on gas-fired heating units and boilers, with or without intermittent pilot ignition, in place of the regular gas valve.

The majority of problems with these valves is electrical rather than mechanical. The redundant gas valve has two electrically controlled internal shutoff valves in series with each other. Before gas can flow to the main burner, both of these valves must be open. However, only one needs to close to stop the flow of gas.

Safety, therefore, is accomplished. It is almost impossible for both of these individual valves to stick in the open position when either the thermostat or the limit control opens. The standard main gas valve does not provide this safety.

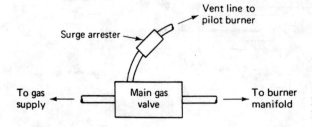

Figure 2–136. Surge arrester in vent line.

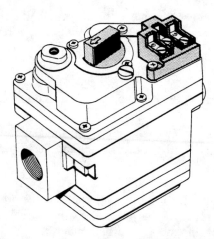

Figure 2–137.   Redundant gas valve.

One make of redundant gas valve uses an instant-acting solenoid valve for controlling the flow of gas at the valve outlet and a time delay valve on the inlet of the valve. The time delay on the bimetal valve is about 10 seconds, thus providing a time delay on startup of the main burner.

The redundant gas valve used on furnaces with standing pilots is similar to the standard type of gas valve except that an internal heat motor valve is included. There are two 24-V coils used in these valves (see Fig. 2–138). One coil is shown as a solenoid valve and the other is shown as a resistor. They are parallel electrically and both are in series with the limit switch and the thermostat. The resistor is for the heat motor that operates the second or redundant gas valve.

These valves have a third internal valve, which is the 100% safety shutoff valve. It is energized with the electricity (millivolts) that is produced by the thermocouple. Should the pilot flame not heat the thermocouple sufficiently, the 100% safety shutoff will close, stopping the flow of gas to the main burner.

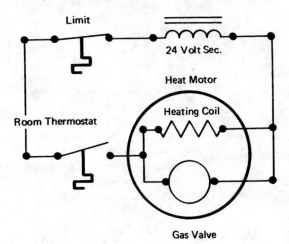

Figure 2–138.   Two 24-V coils in one valve.

When the contacts in either the thermostat or the limit switch open, the solenoid instantly closes. The heat motor valve, however, takes a few seconds to cool down and close the valve. When the standing pilot valve is to be tested, use the following steps:

1. When the pilot is burning, the open-circuit thermocouple test should show a minimum of 21 mV.
2. When the thermostat is calling for heat and the limit switch is closed, 24 V of electricity should be at both the solenoid coil and the heating coil of the heat motor.

If proper voltage is indicated in both of these tests, the problem is not in the electrical circuit. At this point check for gas at the valve inlet. If gas is present, the valve has a mechanical problem and should be replaced.

When an intermittent pilot ignition system is used, each time the thermostat demands heat pilot gas will be supplied and lighted electrically. The redundant gas valve will then open and admit gas to the main burner. If the pilot gas should not be lighted within the specified amount of time, the system will shut down and will lock out on safety.

The valve used on this type of system has two solenoid valves wound on the same core (see Fig. 2–139). When the thermostat demands heat, an electrical circuit is completed to both coils and to the pilot gas ignitor. The pilot gas ignitor will

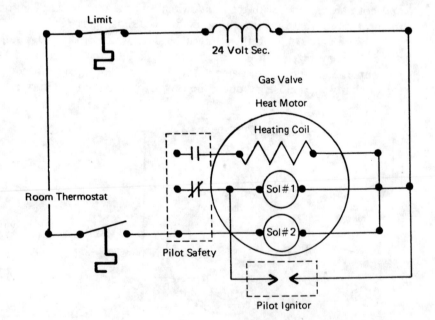

Figure 2–139.   Two solenoid coils on one core.

direct a high-voltage spark across the pilot, igniting the pilot gas. The coil in the pilot safety circuit, which consists of a set of normally open and a set of normally closed contacts, is then energized by the millivolts produced by the heated thermocouple.

When the pilot flame has been established, the millivoltage generated by the thermocouple will cause the contacts in the pilot safety to change position. At this point, the solenoid coil 1 and the pilot gas ignitor are deenergized. The coil in solenoid 2 will remain energized until the thermostat is satisfied. In this type of valve, both solenoids must be energized for the valve to open, but only one is required to keep the valve open.

As the contacts in the pilot safety change positions, an electrical circuit to the main gas valve is completed. The main gas valve is operated by the heat motor valve. It will be approximately 30 seconds before gas is admitted to the main burner.

For gas to flow to the main burner, both solenoids must be energized when the thermostat demands heat. The pilot flame must be established before the main gas valve will open. The two solenoids being wound on the same core prevent cycling of the main burner because of sudden opening or closing of the thermostat contacts or a momentary electrical power failure. Should either of these conditions occur, solenoid 2 will close and will not be energized again until the thermocouple has cooled down and it resets the circuit.

When troubleshooting this valve for electrical troubles, use the following steps:

1. With the pilot lit, there should be 24 V to the main gas valve heater coil.
2. There should be 24 V to solenoid coil 2.
3. When these conditions are met and still no gas is admitted to the main burner, the valve is defective and must be replaced.

2-55 PRESSURE REGULATORS

Natural gas furnaces are designed to operate with 3½ to 4 in. of water-column pressure and LP (liquefied petroleum gas) furnaces with 11 in. of water-column pressure in the furnace manifold. The purpose of the regulator is to maintain these pressures. Pressure regulators are in the same housing as the main gas valve (see Fig. 2-134). Except on LP gas furnaces, the regulator is located at the storage tank. To check the outlet pressure of the regulator, connect a manometer or a manifold pressure gauge to the manifold (see Fig. 2-140). To make any adjustments, remove the cap and insert a screwdriver inside to the adjustment screw. Turn the adjusting screw in to increase or out to decrease the outlet pressure.

A. Large-Capacity Units

On some large-capacity units, a step-opening regulator is used. Step-opening regulators are used to prevent large amounts of gas being input to the combustion area

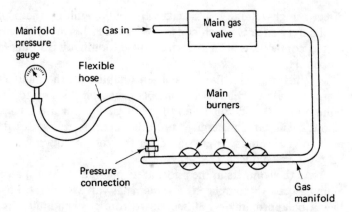

Figure 2-140. Checking manifold pressure on a gas furnace.

all at once. Initially, these regulators open to about 80% of their full capacity, then open fully after the ignition stage has been accomplished. When erratic operation is experienced with these valves, replace the valve rather than make repairs.

B. Unstable Gas Supply

When an unstable gas supply is experienced, a two-stage pressure regulator should be installed in place of the original regulator. An unstable gas supply may be detected by wide variations in the gas manifold pressure on the furnace. A two-stage pressure regulator will smooth out these pulsations and will permit a more even flame.

2-56 HEAT EXCHANGERS

The heat exchanger in a gas-fired furnace acts as the combustion area, the heat transfer medium, and the flue passages (see Fig. 2-141). The heat exchanger is some-

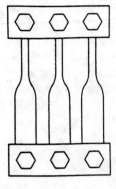

Figure 2-141. Gas furnace heat exchanger.

times called the "heart" of the heating system. The most common problems that occur with a heat exchanger are (1) restrictions, and (2) cracks or openings in the metal.

A. Restricted Flue Passages

Restrictions in the flue passages of a heat exchanger upset the draft through the furnace and interfere with the combustion process. These restrictions must be removed to restore normal operation. To remove restrictions the first step is to remove the main burners and the draft diverter. The top and bottom openings are now clear to insert a straightened-out coat hanger (see Fig. 2–142). Scrape each side of the flue passage from front to back several times to loosen any rust particles and soot. The loosened rust and soot will fall to the bottom of the combustion zone and must be removed. After the rust and soot are removed from the heat exchanger, clean the burners and reinstall them. Light the furnace and adjust the main burners to obtain good combustion. The heat exchanger should next be checked for cracks (see Section B).

B. Cracked Heat Exchangers

Heat exchangers that are cracked are hazardous because of possible fire and carbon monoxide poisoning. A cracked heat exchanger allows the recirculating air and the products of combustion to mix. When the fan is running, a pressure is built up in the duct system that forces air into the combustion zone and causes the flame to be blown around. When the crack is large enough, the flame will be blown out the front of the combustion zone. To make a more positive check, turn off the gas to the furnace and place a smoke bomb in the combustion zone. Place the bomb first in one section and then in another while the fan is running. The smoke will be blown out of the burner opening. A cracked heat exchanger should be replaced, never repaired. A heat exchanger that has cracked once will probably crack again.

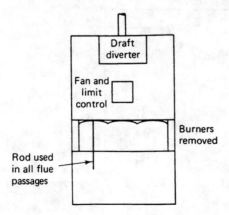

Figure 2–142. Cleaning a plugged heat exchanger.

2-57 GAS VENTING

The purpose of a venting system is to provide an escape of the products of combustion to the atmosphere. As the products of combustion leave the furnace heat exchanger, they enter the draft diverter (see Fig. 2–143). In the draft diverter the products of combustion are mixed with air from the room and diluted. The mixture then passes through the vent pipe to the outside atmosphere. When the vent pipe fails to remove the products of combustion, they are directed from the draft diverter into the building, thus creating odors in the building and upsetting the combustion process. This condition can also be caused when an exhaust fan is installed in the equipment room and is not supplied with enough makeup air. To check the vent action, hold a lighted match close to the draft diverter opening while the furnace is operating (see Fig. 2–144). If the vent is operating properly, the match flame will be drawn toward the draft diverter. If the vent is not operating properly, the match flame will be blown away from the draft diverter and may be snuffed out. To correct, remove the obstruction and stop the down-draft condition by installing a new vent cap or by locating the vent termination point out of the high-static-pressure area. When an exhaust fan is used, be sure that enough makeup air is provided.

2-58 LIMIT CONTROL

The limit control is a safety device used to interrupt electrical power to the main gas valve when an overheated condition occurs. It is a bimetal temperature-sensing switch with normally closed contacts that open at approximately 200°F (93°C). To check the limit control, insert a thermometer in the recirculating airstream as close to the sensing element as possible. Reduce the airflow through the furnace and ob-

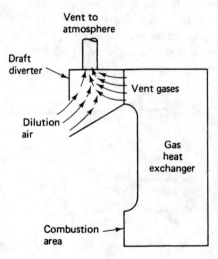

Vent to atmosphere

Draft diverter

Vent gases

Dilution air

Gas heat exchanger

Combustion area

Figure 2-143. Vent gases and dilution air mixing in draft diverter.

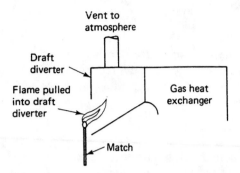

Figure 2–144. Checking vent action with a wooden match.

serve the thermometer when the main burner flame goes off. If this temperature is about 200°F (93°C), the limit control is functioning normally. The trouble is something else. If the temperature indicated is not quite 200°F (93°C), adjust the control; make only minor adjustments on the limit control. If the temperature is off by as much as 20°F (9.4°C), replace the limit control. Be sure that the replacement control has the same-length sensing element (see Section 4–4).

2–59 END PLAY IN MOTOR

Electric motors are equipped with fiber washers, or spacers, on the shaft so that alignment of the rotor and the magnetic field may be possible (see Fig. 2–145). These washers or spacers become worn and allow the rotor to float back and forth in the bearings. This floating movement will cause a noise indicating that repair is needed. This repair is usually done in an electric motor repair shop. Therefore, the motor must be removed and delivered. When replacing the motor, follow the manufacturer's recommendations for the proper procedures.

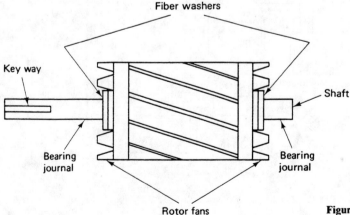

Figure 2–145. Fiber washers on shaft.

2-60 CHECK VALVES

Check valves are used on heat pump systems to help in directing the flow of refrigerant through the proper path for the desired cycle: heating, cooling, or defrosting (see Fig. 2-146). The check valve will allow the passage of refrigerant either around the flow control device or through the flow control device. Use the following suggestions to check the operation of a check valve.

1. A check valve sticking in the closed position in the outdoor section during the cooling cycle will cause the suction pressure to be low. The superheat will be high on the indoor coil.
2. A check valve sticking open in the outdoor section during the cooling cycle will cause high suction pressure, flooding of the compressor, and low superheat on the indoor coil.
3. A check valve sticking closed in the outdoor section during the heating cycle will cause the suction pressure to be low and the superheat to be high on the outdoor coil.
4. A check valve sticking open in the outdoor section during the heating cycle will cause a high suction pressure, flooding of the compressor, and a low superheat on the outdoor coil.

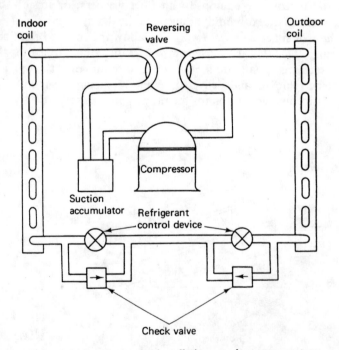

Figure 2-146. Check valve installation on a heat pump system.

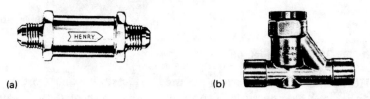

Figure 2-147. Check valves: (a) ball check; (b) swing check. (Courtesy of Henry Valve Co.)

If any of the conditions above are found, the check valve should be replaced. Be sure to purge the refrigerant charge or pump the system down to prevent personal injury. Be sure to install the check valve with the arrow pointing in the direction of refrigerant flow at that point (see Fig. 2-147).

2-61 REVERSING VALVE

The purpose of the reversing valve is to reverse the refrigerant direction on a heat pump, and reverse cycle refrigeration systems. Some systems energize the coil on heating and some energize the coil on cooling. Therefore, the system in question will determine that energized condition of the coil. Any troubles that occur in a heat pump system that will affect the normal operating pressures could possibly affect proper shifting of the valve. Some examples of these problems are a refrigerant leak in the system resulting in a shortage of refrigerant, a compressor that is not operating to full capacity, a defective check valve, a damaged valve, or a defective electrical system. Any of these conditions will indicate an apparent malfunctioning valve. The following checks should be made on the system and its components before attempting to diagnose any valve troubles by using the "touch test" method.

1. Inspect the electrical system. This is best done by having the system in operation so that the reversing valve solenoid is energized. While the unit is operating, remove the lock nut from the solenoid cover. Pull the solenoid coil partway off the stem. There should be a magnetic force trying to hold the coil on the stem. Be sure that electric power is applied to the solenoid. If the coil is moved off the stem, the valve will return to the deenergized position, which will be indicated by a clicking sound. A clicking sound will also be heard when the coil is replaced on the stem.
2. Inspect the reversing valve for damage. Check the solenoid coil for deep scratches, dents, and cracks.
3. Check the operation of the system against the equipment manufacturer's recommendations. Make any corrections indicated.

When replacing the reversing valve, the refrigerant will need to be purged. Follow the manufacturer's recommendations when installing a new valve. Evacuate the system, install new filter-driers, and recharge the system according to the manufacturer's recommendations.

2-62   DEFROST CONTROLS

Heat pump defrost controls are used to detect ice and frost on the outdoor coil during the heating cycle. The most popular type of defrost control is the time–temperature method. This method means that both the time and temperature sections must demand defrost before the defrost cycle can be initiated. When the defrost cycle is initiated, the reversing valve changes and the outdoor fan motor stops running. The clock motor operates only when the compressor is running. The timer can be set to initiate the defrost cycle on 30-, 60-, and 90-minute intervals. If it is found that the 90-minute cycle allows too much frost on the outdoor coil, then the timer can be adjusted so that less frost accumulates on the coil before the defrost cycle is initiated. The defrost can be initiated only by demand from both the timer and the temperature sections of the defrost control. The defrost control temperature will initiate the defrost cycle at approximately 30°F (-1.1°C).

     The timer motor is operated by 230 V and operates only when the compressor is operating. The temperature portion of the control is attached to a return bend on the outdoor coil (see Fig. 2–148). The bulb is insulated to ensure that the ambient air does not affect the bulb. To check a time–temperature defrost control, be sure that the outdoor coil is cold enough to need defrosting; then use the following steps:

1. Place a jumper between terminals 2 and 3 on the clock timer (see Fig. 2–149). If the reversing valve changes, the clock is not operating. Check to make certain that power is supplied to the clock motor terminals. If power is found, the motor is defective and must be replaced.

2. If the reversing valve does not change, jump the terminals of the defrost thermostat while the jumper remains on terminals 3 and 4. If the reversing valve changes, the defrost control is defective and must be replaced.

     If the additional jumper does not cause the reversing valve to change, the trouble is not in the defrost control. The reversing valve solenoid coil, the reversing valve, and the defrost relay must be checked to find the trouble. Be sure that the specified voltage is applied to these controls.

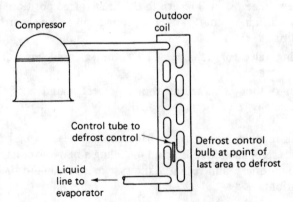

Figure 2–148.   Installation of temperature bulb for a defrost control.

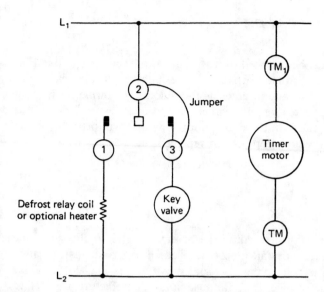

Figure 2-149.   Jumper between terminals 2 and 3 on defrost timer.

2-63 AIR BYPASSING COIL

Air bypassing a coil will reduce the efficiency of the unit (see Fig. 2-150). The greater the amount of air bypassing the coil, the greater the reduction in efficiency. The opening through which the air bypasses must be closed so that peak efficiency can be obtained. Air bypassing an evaporating coil causes low suction pressure. Air bypassing a condensing coil causes high discharge pressure.

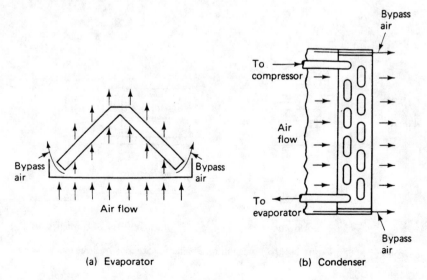

(a) Evaporator   (b) Condenser

Figure 2-150.   Air bypassing coils.

2-64 AUXILIARY HEAT STRIP LOCATION

The auxiliary heat strips must be installed downstream of the indoor coil (see Fig. 2-151). If the heat strips are installed upstream of the indoor coil, the heat pump will act like a reheat system. This hot air entering the indoor coil during the heating cycle will cause the discharge pressure and temperature to be excessively high, thus causing a high compression ratio on the compressor, one that will probably cause permanent damage and will require compressor replacement.

2-65 DEFROST RELAY

The purpose of the defrost relay is to cause the reversing valve to change to the cooling position, stop the outdoor fan motor, and energize the auxiliary heaters during the defrost cycle. The coil voltage is usually 240 V (see Fig. 2-152). This relay is energized by the defrost control. Because some manufacturers prefer to energize the reversing valve during the heating cycle and some during the cooling cycle, the wiring diagram for the particular unit in question should be used.

If the relay is energized and a click is heard when the defrost control demands defrost but the system does not shift, check the voltage across each set of contacts with a voltmeter. Open contacts will show full line voltage across them. Closed contacts will show no line voltage across them. If conditions other than these are found, replace the relay.

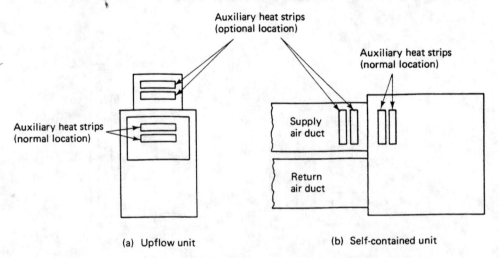

(a) Upflow unit   (b) Self-contained unit

Figure 2-151.   Location of auxiliary heat strips on heat pump systems.

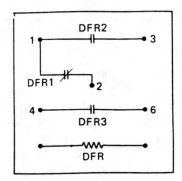

Figure 2-152.   Internal schematic of a defrost relay.

2-66 SHORTED FLAME DETECTOR CIRCUIT

The purpose of the flame detector is to determine whether or not a satisfactory flame has been established in the firebox of an oil burner unit. If a satisfactory flame has not been established in approximately 60 seconds, the flame detector will shut down the oil burner. To check and determine whether the flame detector is defective, improperly positioned, or dirty, use the following procedure:

1. Remove the cad cell from the flame detector socket assembly. To remove cell, push in and turn counterclockwise. If there is any soot accumulation on the face, remove it with a soft cloth. Replace the cell in the socket assembly and start the burner. If the burner runs and does not lock out, the cad cell was dirty.

2. If the primary control is locked out on safety or if the cad cell face was found clean (step 1), install a new cad cell and start the burner. If the burner runs and does not lock out, the cad cell was defective.

3. If the primary control locks out on safety (in step 2), the flame detector is not properly positioned, or there is an open circuit between the contacts on the cad cell and the socket assembly. Follow the manufacturer's recommendations in repositioning the flame detector.

4. If the burner will not start, the flame detector leads of the cad cell may be shorted, or there could be a false light affecting the cad cell. Use the following procedure for testing:
   a. Remove one of the leads from FD-FD terminals.
   b. Turn on the electricity. If the burner starts, the flame detector is defective and must be replaced.
   c. If the burner does not start, the trouble is elsewhere and other checks must be made.

The proper application, location, and mounting of the flame detector units are determined by the equipment manufacturer. Every service technician should be familiar with the following factors, which are the basis for flame detector location:

1. The flame detector must be located so that its face temperature never exceeds 140°F (60°C). Therefore, it must be mounted in a cool location.
2. The flame detector must have a direct view of the burner flame. On some installations, it might be necessary to drill holes in static disks, and so on, to provide a clear view for the flame detector.

The flame detector should not be adjusted to react to reflected light because the amount of reflection changes after the burner is in operation a short period of time. Some equipment manufacturers blacken the inside of the firebox to simulate actual field conditions when designing new installations.

2-67 SHORTED FLAME DETECTOR LEADS

Shorted leads will bypass the flame detector unit and will prevent the burner from starting. The shorted leads must be found and must be separated and insulated. Precautions should be made to prevent the return of this problem. It may be necessary to install new wiring, using rubber grommets and other protection at points where wear of the wiring may occur.

2-68 FRICTION CLUTCH

The purpose of the friction clutch in a flame detector is to prevent damage to the detector due to the stack temperature. When a flame has been established, the rise in stack temperature causes the thermal element on the stack control to expand, moving the stack switch arm down, closing the stack switch, and shutting out the safety switch heater. The stack switch arm moves down to the lower (hot) stop. The additional expansion of the thermal element is absorbed by the friction clutch. After the thermostat stops the burner, the drop in stack temperature causes the thermal element to contract, opening the stack switch and returning the stack arm to the starting position. Any further contracting of the thermal element is absorbed by the friction clutch. A faulty friction clutch will prevent both the normal operation of the flame detector and starting of the oil burner. The detector, or the element, must be replaced to correct the problem.

2-69 HOT AND COLD FLAME DETECTOR CONTACTS

The hot flame detector contacts are used to shunt out the heater element in the safety switch, and the cold contacts are used to control the ignition process. In operation, when the thermostat calls for heat, the relay coils are energized and the right-hand relay pulls in. The ignition is started when contact 15 closes. The left-hand relay is energized when contacts 9, 10, and 11 are energized. Contact 16 starts the oil burner motor. The hot contact 8 is closed, shunting out the heater in the safety switch as the stack temperature rises. A further increase in the stack temperature causes contacts 7 and 6 to break, in that order, thus interrupting the circuit to the right-hand relay, dropping it out, and shutting off the ignition. The burner is now operating normally.

If any one of these contacts become dirty, pitted, stuck, or inoperative, it must be serviced before normal operations will resume. Dirty contacts may be cleaned by placing a business card between them and pulling it back and forth until the contacts are clean. Pitted contacts may be filed lightly with a contact file. If the contacts are severely pitted, the control should be replaced or the trouble will reoccur. Stuck or inoperative contacts require that the control be replaced.

2-70 FLAME DETECTOR BIMETAL

The purpose of a bimetal in a flame detector is to sense the stack temperature and cause the proper relays and contacts to act accordingly. Should this bimetal become covered with soot or warped, the flame detector will not respond properly. If the bimetal is covered with soot, the safety switch will lock the burner out before the detector has time to switch into the normal operating position. And if the bimetal is warped, it will not permit proper operation of the flame detector and must be replaced. It is generally best to replace the complete control rather than simply replacing the bimetal. A dirty bimetal must be cleaned in order to resume normal operation. A good cleaning solvent such as kerosene or gasoline may be used for cleaning. Use caution not to twist or warp the bimetal.

2-71 PRIMARY CONTROL

The purpose of the primary control is to control operation of the oil burner and the ignition procedure. After the control has been properly installed and wired, the burner should operate as follows. On demand from the thermostat, the combustion relay is closed, starting the oil burner motor and the ignition process. The safety-switch heater is also energized at the same time. When a predetermined period of time has passed, the ignition timer stops the ignition process. If the safety-switch heater is not shunted out by the stack switch, a safety shutdown will occur. To restart the unit, reset the safety switch.

Any malfunction in the primary control could be due to a defective internal circuit or dirty combustion relay contacts. A defective internal circuit will require a new primary control. Dirty combustion relay contacts may be cleaned by moving a business card back and forth between them. File pitted contacts lightly or the contacts will be ruined, requiring replacement of the control.

Air leaking into the vent pipe around the flame detector mount will keep the bimetal cool and will cause the control to lock out on safety. Sealing any leaks here should be done with asbestos-type gaskets because of the high temperatures encountered (see Fig. 2–153).

2–72 OIL BURNER BLOWER WHEEL

The purpose of the blower on an oil burner is to furnish the combustion air to the flame (see Fig. 2–154). A binding blower wheel will reduce or completely stop the combustion air. This binding could be due to a loose setscrew, allowing the blower to get out of alignment, or it may be due to dirty or worn bearings. To test for a binding blower wheel, simply rotate the blower by hand. If the blower rubs on the fan scroll, it needs adjustment to clear all components. If the blower comes to a sudden stop after manual rotation rather than coasting to a stop, the trouble is in the bearings. The bearings should be lubricated and tested for wear. If worn, they should be replaced.

2–73 FUEL PUMP

The purpose of the fuel pump is to supply the fuel oil to the burner with sufficient pressure to atomize the fuel for combustion. There are basically two different pressures used to atomize the fuel: high pressure [100 psig (687 kPa)] and low pressure [15 psig (204.03 kPa)]. A fuel pump is adjustable, within limits, to provide the prescribed pressure for the burner (see Fig. 2–155). The method and location of the pressure adjustment will vary with the type and make of the pump. Refer to the manufacturer's specifications for the pump in question. These pumps are usually

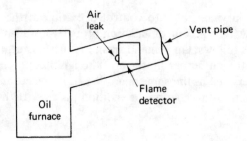

Figure 2-153. Air leak in vent pipe.

Figure 2-154. Location of oil burner blower wheel. (Courtesy of Lennox Industries.)

very trouble-free. However, occasionally a bearing will seize, rendering a pump inoperative. In such cases, the pump must be replaced. Be sure to follow the manufacturer's installation and adjustment recommendation. Also, occasionally a drive belt or a drive coupling will break and need to be replaced. Fuel pumps are usually equipped with strainers to prevent the entrance of foreign matter into the pump. These strainers occasionally become plugged and will not allow the pump to produce the desired pressure. Plugged strainers require cleaning. Cleaning the strainers is a

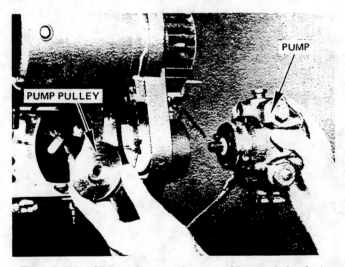

Figure 2-155. Oil burner pump. (Courtesy of Lennox Industries.)

relatively simple task. Stop the pump, close off the fuel-line valve, disconnect the inlet fuel line, and remove the strainer. Clean the strainer with gasoline or some other suitable solvent. Reinstall the strainer, reconnect the fuel line, and bleed any air from the line by opening the fuel-line valve and allowing a small amount of fuel to escape before tightening the line connection. Place the unit back in operation.

2-74 OIL BURNER ADJUSTMENTS

To set an oil burner properly, a CO_2 analyzer and a smoke gun must be used. Do not set oil burners by eye or estimation. When instruments are not used, the burner will be set with too high CO_2, and smoke will result, which is usually accompanied by a noisy fire and carbon buildup in the heat exchanger. A proper setting with instruments will result in a quiet and clean flame at 8 to 10% CO_2, with zero to a slight trace of smoke.

Before making any final burner adjustments, allow the burner to operate continuously for 5 to 10 minutes. This operation will purge the fuel lines and will level out the combustion process. Take the readings and make the adjustments as follows:

1. Make sure that the inspection door is closed tightly and that any fitting joints between the furnace and the point where the CO_2 and smoke readings are taken are tightly sealed or taped. An air leak at the inspection door or a fitting will cause a false CO_2 reading because of diluting the flue gases with air.

2. Punch a $5/16$-in. (7.94-mm)-diameter hole in the flue outlet between the furnace and the draft control. The draft readings, the CO_2, and the smoke test should be taken at this point.

3. Adjust the barometric draft control in the stack for the correct draft (see Fig. 2-156). The draft should be measured with a draft gauge and set for 0.03 to 0.035 in. water column (see Fig. 2-157).

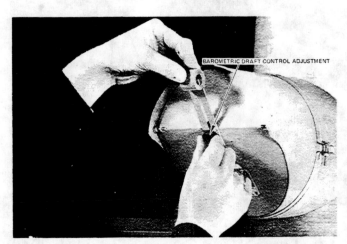

BAROMETRIC DRAFT CONTROL ADJUSTMENT

Figure 2-156. Barometer draft control. (Courtesy of Lennox Industries.)

DRAFT GAUGE

Figure 2-157. Measuring draft on oil burner. (Courtesy of Lennox Industries.)

4. Loosen the air control locking screw and rotate the air control cover until a clean flame is produced (see Fig. 2-158).

5. Take a CO_2 reading at the service opening in the stack using a CO_2 indicator (see Fig. 2-159). Follow the instructions with the CO_2 indicator. If the CO_2 reading is between 8 and 10%, the setting is correct. If not, rotate the air control cover and recheck until the CO_2 readings fall within the 8 to 10% range. Tighten the air control locking screw.

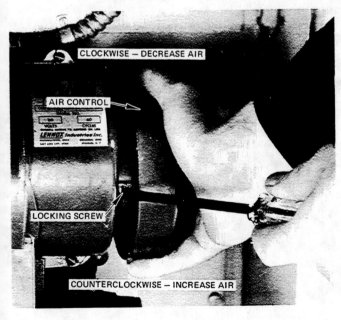

CLOCKWISE — DECREASE AIR

AIR CONTROL

LOCKING SCREW

COUNTERCLOCKWISE — INCREASE AIR

Figure 2-158. Adjusting an oil burner air control cover. (Courtesy of Lennox Industries.)

CO₂ ANALYZER

Figure 2-159. Checking stack CO_2 content. (Courtesy of Lennox Industries.)

6. Take a smoke reading in the same service hole used for taking the CO_2 reading (see Fig. 2-160). Use a standard smoke tester such as the Bacharach True Spot Tester. The smoke reading obtained at an 8 to 10% reading will generally be between zero and a number 1 spot. The smoke reading should never be more than a number 1 spot. If the smoke test provides greater than a number 1 spot, it could be caused by a poor nozzle or by air leakage at the inspection door or fitting. Occasionally, it could be caused by a difference in oil or by some unusual condition created during the original installation. Rotate the combustion air control cover until a number 1 spot or less is obtained; then

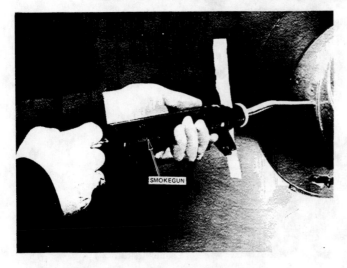

SMOKEGUN

Figure 2-160. Taking a smoke reading of an oil burner. (Courtesy of Lennox Industries.)

recheck the CO_2 to make sure that it is 8 to 10%. If the CO_2 is less than 8%, look for air leakage, a bad nozzle, or improper settings of the burner gun assembly. Air leakage can often be determined by taking a CO_2 reading both at the stack and over the flame. If the stack CO_2 is more than ½% below the CO_2 reading taken over the flame, it indicates an excess air leakage. Any air leakage will lower the furnace efficiently and should be found and corrected.

2-75 OIL TANKS

Oil tanks are used to store the fuel until it is needed. These tanks may be installed inside, outside above ground, or outside below ground level. When fuel oil has been stored for sometime, condensation will occur. This condensation will form in the bottom of the tank in the form of water. This water may be drawn into the oil burner and can upset the combustion process and at times cause a flameout. When this condition exists, the water must be removed from the tank. This may be accomplished by lowering a line into the tank through the oil and into the water collected at the bottom of the tank (see Fig. 2-161). A pump is then connected to the line and the water pumped out. Sometimes it may be necessary to drain the tank completely, to remove all the water.

2-76 FUEL OIL LINES

Fuel oil lines are used to move the oil from the tank to the oil burner. There are both one-pipe and two-pipe systems in use. The number of pipes is dependent on the type of fuel unit used. There is usually a filter installed in the oil supply line to prevent any foreign matter entering the fuel unit (see Fig. 2-162). These fuel lines should be protected to prevent damage such as a kinked line, a dented line, or a leak.

    The fuel supply line is used to move the oil from the tank to the burner and usually has a reduced pressure inside it. Should this line or filter become clogged, the fuel flow will be reduced, affecting operation of the burner. The kink must be removed or a new filter installed. A leak in the supply line will allow air to enter

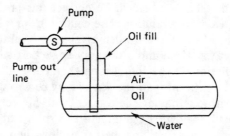

Figure 2-161. Pumping water out of an oil storage tank.

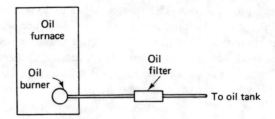

Figure 2-162. Oil filter location.

the system and will cause air blockage of the oil pump. The leak must be found and repaired. Usually, a leak will be indicated by a slight trace of oil around the leak. The air must be purged from the system before normal operation can be resumed.

The fuel return line carries any excess oil back to the oil tank. This line is under pump pressure. Therefore, a leak in this line will allow the fuel oil to escape, causing a mess. A restriction will cause a reduced flow to the tank, thereby putting the rest of the system under excessive pressure and affecting normal operation. Any leaks or restrictions in this line must also be removed.

2-77 OIL BURNER NOZZLES

One of the functions of a nozzle is to atomize the fuel to break it up into tiny droplets that can vaporize in a short period of time. The atomizing nozzle performs three basic and vital functions for an oil burner: atomizing, metering, and patterning. To accomplish proper atomization of the fuel, the nozzle must be designed for the pressure being pumped by the oil pump. If proper atomization is not being obtained and the pump is delivering the prescribed pressure, the trouble is in the nozzle.

The size of the orifice determines the amount of fuel being delivered to a heating unit. Too small an orifice will not deliver enough fuel, whereas too large an orifice will deliver too much fuel. A clogged orifice will not deliver enough fuel and the flame will be one-sided. To correct this problem, remove the orifice and clean it with a suitable solution. Use caution to prevent damage to the orifice.

The nozzle is expected to deliver atomized oil to the combustion chamber in a uniform spray pattern and at a spray angle best suited to the requirements of a specific installation. If the combustion chamber is almost but not completely filled with flame, the patterning of the nozzle may be wrong. In this case the nozzle must be replaced with one having the correct patterning characteristics. When replacing a nozzle, consult the orifice manufacturer's specification chart to obtain the correct nozzle for the installation. It is best to follow the recommendations of the equipment manufacturer.

2-78 NOZZLE FILTER

The nozzle filter or strainer is designed to prevent dirt or other foreign matter from getting into the nozzle and clogging its passages. Nozzle manufacturers specify the type and size based on protecting the smallest passage in the nozzle slots. When the nozzle filter becomes clogged, the flow of fuel is reduced, and underfiring of the unit results. A clogged nozzle filter requires that the nozzle be replaced with one having the proper characteristics, as outlined in Section 2–77.

2-79 IGNITION TRANSFORMER

The ignition transformer, commonly known as a step-up transformer, is used to step up the voltage from line voltage to 5000V or more to jump the gap at the end of the ignition electrode (see Fig. 2–163). Ignition transformers are quite trouble-free and require only that the primary voltage correspond to that of the transformer, that the electrical connections be clean and tight, and that the secondary terminals be kept clean.

If the secondary terminal post becomes fouled with soot, dust, or other foreign matter, a high-tension spark can easily short-circuit to the transformer case, resulting in no spark at the ignition electrode. A transformer in good condition should always cause a spark to jump a gap of not less than ¼ in. (6.4 mm). The transformer case must be securely grounded to provide a path for the current to flow. The spark should be a bright blue color.

To check an ignition transformer, first check to see that voltage is supplied to the primary side. If voltage is found on the primary side, check to see if there is voltage at the secondary terminals. *Caution:* To prevent meter damage, be sure that the meter used is positioned on a high-enough scale. If no voltage is found on the secondary terminals, and there is voltage to the primary side, the transformer is bad and must be replaced.

If secondary voltage is found and there is no sparking at the electrode ends, check the transformer ground wire and make certain that it is clean and tight. Next check the electrodes for cleanliness, gap, and loose connections or broken insulators (see Fig. 2–164). Clean and gap the ignition electrodes and replace any defective electrode leads or defective insulators. The electrode gap should be set according to the manufacturer's specifications. Clean and tighten all electrical connections to the ignition circuit.

2-80 FUEL OIL GRADE

Fuel oil is refined in several different grades, ranging from No. 1 to No. 6 and omitting No. 3. The higher the number, the thicker the oil. Therefore, Nos. 1 and 2 are the most popular grades for domestic and light commercial use. Occasionally,

Figure 2-163. Component location of oil burner. (Courtesy of Lennox Industries.)

in cold weather where the tank is not protected from the low temperatures, No. 2 fuel oil will become too thick for the pump to move properly. When this situation occurs, drain the tank and refill it with No. 1 oil. If only a small amount of No. 2 is in the tank, it may not be necessary to drain the tank because the No. 1 oil will reduce the viscosity, allowing the No. 2 oil to flow properly. In such cases it becomes necessary to drain only the fuel supply line.

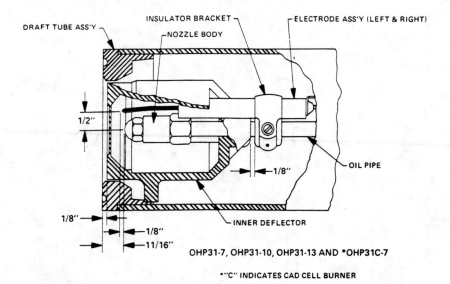

OHP31-7, OHP31-10, OHP31-13 AND *OHP31C-7

*"C" INDICATES CAD CELL BURNER

Figure 2-164.   Location of ignition electrodes. (Courtesy of Lennox Industries.)

3 Startup Procedures

3-1 GAS HEATING

1. Close off all gas valves.
2. Wait approximately 5 minutes for any unburned gas to escape from the combustion area.
3. Open the gas cock in the gas line.
4. Turn the off–on pilot knob to the "pilot" position.
5. Strike a match and hold the flame by the pilot burner.
6. Depress the pilot knob (see Fig. 3-1).
7. Hold the pilot knob in for approximately 1 minute after the flame has been lit.
8. Release the pilot knob.
9. Turn the room thermostat down below room temperature or turn off electrical power to the furnace.
10. Turn the gas valve knob to the "on" position.
11. Turn on the electrical power or raise the temperature setting above room temperature.
12. Check the flame on the main burners. Adjust the primary air shutter if there is any yellow in the flame. If the burner cannot be adjusted properly, close off the gas valve, and remove and clean the burners. Reinstall the burners, complete the steps above, and adjust the primary air.

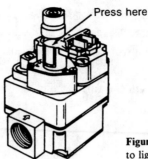

Figure 3–1. Pressing knob on gas valve to light pilot.

13. Check for carbon monoxide in the flue gases inside the draft diverter. Make any adjustments indicated (see Fig. 3–2).
14. Check the fan "on" and "off" temperatures and the limit "off" temperature with a thermometer inserted into the circulating airstream as close as possible to the sensing element (see Fig. 3–3).
15. Check the heat anticipator.
16. Turn off the electrical power to the furnace after the fan has stopped running.
17. Clean or replace the air filter.

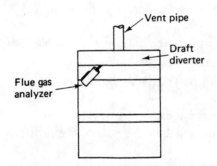

Figure 3–2. Checking flue gases.

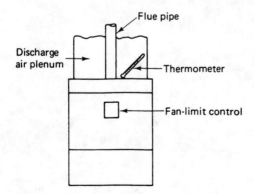

Figure 3–3. Checking fan-limit control operation.

18. Check the condition of all bearings. Repair or replace as required.
19. Lubricate all bearings requiring this service.
20. Check the belt tension. Replace or adjust as required (see Fig. 3–4).
21. Check the calibration of the thermostat. Calibrate as required.
22. Clean and vacuum the furnace and the area around the furnace.
23. Replace all covers.

3-2 HEAT PUMP HEATING

1. Check to be sure that the electricity has been on to the outdoor unit for at least 24 hours.
2. Check the outdoor coil and clean if necessary.
3. Check all bearings; make necessary repairs.
4. Lubricate all bearings requiring this service.
5. Check the condition and tension of all belts. Replace or adjust as required (see Fig. 3–4).
6. Check the condition of the compressor contactor contacts. Replace if necessary.
7. Check the tightness of all electrical connections. Tighten loose connections.
8. Check and repair burnt or frayed wiring.
9. Set the thermostat for heating, and raise the temperature setting above room temperature.
10. Check the suction and discharge pressures.
11. Check the refrigerant charge in the system. If short, repair the leak and add refrigerant.
12. Check the amperage draw to all motors.

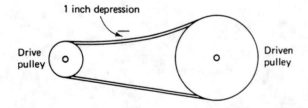

Figure 3-4. Checking belt tension.

13. Check the temperature rise on the indoor coil. The temperature rise should be approximately 30°F (16.67°C) (see Fig. 3-5).

14. Check the thermostat calibration. Calibrate if necessary.

15. Clean or replace the air filter.

16. Clean and vacuum the indoor unit and the area around the unit.

17. Replace all covers.

3-3 ELECTRIC HEATING

1. Turn off all electric power to the unit.

2. Visually check for any burned or bad wiring and replace as required.

3. Tighten all electrical connections.

4. Turn on the electric power.

5. Set the room thermostat well above room temperature to ensure that all stages are demanding.

6. Check the voltage drop on all elements after all relays are closed.

7. Check the amperage to all elements. Replace any bad elements.

8. Check the setting of the heat anticipator. Adjust as required.

9. Set the thermostat selector below room temperature.

10. After the fan has stopped, turn off all electricity.

11. Check the condition of all bearings.

12. Lubricate all bearings requiring this service.

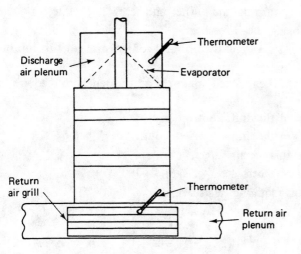

Figure 3-5. Checking temperature rise on indoor coil.

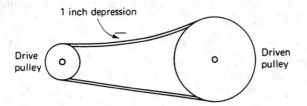

Figure 3–6.   Checking belt tension.

13. Check the condition and tension of all belts. Replace or adjust as required (see Fig. 3–6).
14. Replace or clean the air filter.
15. Clean and vacuum the furnace and the area around the furnace.
16. Replace all covers.
17. Turn on the electricity.

3–4 OIL HEATING

1. Check the oil level in the storage tank.
2. Open the valve in the oil supply line to the burner.
3. Bleed the air from the fuel pump. One-pipe systems must be bled. Two-pipe systems will usually bleed themselves.
4. Turn on the electric switch.
5. Place a can under the bleed port to catch purged oil. Loosen the bleed plug and start the burner. Allow the burner to run until a solid stream of oil is purged from the port. Turn off the burner and screw in the bleed port.
6. Set the room thermostat to call for heat.
7. If the burner does not start, reset the primary control and reset the oil burner motor overload.
8. After the oil burner has been started, make the adjustments outlined in Section 2–74.
9. Check the operation of the draft control.
10. Check the operation of the primary safety control.
11. Clean or replace the furnace air filters.
12. Adjust or replace the fan belt (see Fig. 3–6).
13. Lubricate all bearings requiring this service.

14. Check the fan control settings for accuracy (see Fig. 3-3).
15. Check the limit control setting for accuracy.

4

Standard Service Procedures

4-1 PROCEDURE FOR REPLACING COMPRESSORS

Past experience has demonstrated that after a hermetic motor burnout has occurred, the refrigeration system must be cleaned to remove all contaminants. Without removal of these contaminants, a repeat burnout will occur in a short period of time. Failure to follow minimum cleaning recommendations as quickly as possible will result in an excessive risk of repeat burnout.

Cleaning the system by flushing with Refrigerant-11, or similar refrigerants, has been used in the past under certain conditions and with mixed degrees of success. This flushing method is seldom used today and is not recommended here. When there is liquid water in the system some type of flushing is desirable and this is usually done with Refrigerant-11. This condition occurs when a water-cooled condenser or water chiller has ruptured and water is forced into the system. This water must be removed by whatever means possible.

The first step is to be certain that the compressor is burned. The following list is a general procedure used:

1. When a compressor fails to start, it may appear to be burned out. However, the fault can be in many other areas.
2. All other possibilities should be eliminated before the compressor is condemned. Check for electrical and mechanical misapplication or malfunction.
3. Check the compressor motor for shorted, grounded, and open windings. If none are found, check the winding resistance with a precision ohmmeter to determine if turn-to-turn shorts exist.

If the motor is found burned, the refrigerant should be discharged from the system. The service technician should follow certain safety practices. In addition to the electrical hazards, the service technician should be aware of the dangers of receiving acid burns.

1. When discharging refrigerant (gas or liquid) from the system in which a burn-out has occurred, avoid injury by not allowing the refrigerant to touch the eyes or skin. If the complete charge is to be purged, it should be discharged outside the building. The burned refrigerant vapors will tarnish metal surfaces as well as contaminate the air used for breathing when released inside a building.
2. When it is necessary for the service technician to come in contact with the oil or sludge in a burned-out compressor, rubber gloves should be worn to prevent possible acid burns.

When replacing the compressor in a system that has a burnout, the system must be thoroughly cleaned to remove as many of the contaminants as possible. The following is a list of recommended procedures:

1. On systems of 5 hp and smaller, discharge the refrigerant charge to the outside atmosphere—preferably in the liquid form. On units larger than 5 hp, an attempt to save the refrigerant should be made by valving off the compressor (front-seating the service valves), then purging the refrigerant from the compressor.
2. Install an approved cleanup filter-drier in the suction line at the compressor. On units in which the refrigerant has been saved, this cannot be done until the refrigerant has been pumped from the low side of the system (see Fig. 4-1).
3. Install an approved oversized liquid-line cleanup drier in the liquid line (see Fig. 4-2). On units in which the refrigerant has been saved, this cannot be done until the refrigerant has been pumped from the high side of the system.
4. Check the refrigerant flow control device and clean it thoroughly or replace it.

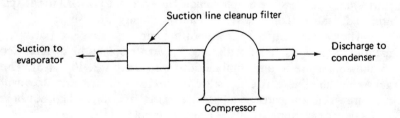

Figure 4-1.   Suction line cleanup filter drier location.

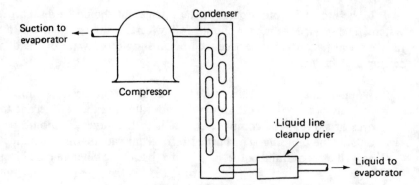

Figure 4–2.   Liquid line cleanup drier location.

5. Remove the inoperative compressor and install the proper replacement. Make all tubing connections leak-tight.

6. Purge the system by blowing refrigerant through the system, forcing the refrigerant completely through all piping and coils.

7. Triple-evacuate the system. The last evacuation should lower the system pressure to 1000 microns or lower, if time permits.

8. Recharge the system and put it back in operation. On systems in which the charge has been saved, add refrigerant to complete the charge.

9. After the system has been in operation for approximately 24 hours, the acid content of the oil and the condition of the cleanup driers and filters should be checked. If acid is found in the oil, noted by discoloration, or the cleanup filters and driers show signs of stoppage, they should be replaced, the oil drained, and a new oil charge put into the system. This procedure should be repeated until a clean system is indicated.

4–2 PROCEDURE FOR USING THE GAUGE MANIFOLD

The gauge manifold is probably the most important tool in the service technicians toolbox. The various "gauges" can be used to check system pressures, to charge refrigerant into the system, to evacuate the system, to add oil to the system, to purge noncondensables from the system, and for many other uses.

The gauge manifold consists of a compound gauge, a pressure gauge, and a manifold that is equipped with hand valves to isolate the different connections or to allow their use in any combination as required (see Fig. 4–3). The ports to the gauge and the line connections are connected so that the gauges will indicate the pressure when connected to a pressure source. Flexible, leakproof hoses are used to make the connections from the gauge manifold to the system.

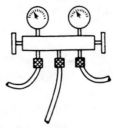

Figure 4-3. Gauge manifold.

One of the most common service functions is hooking the gauges to the system. Care should be taken to prevent contaminants from entering the system. The hoses should be purged by allowing a small amount of refrigerant to escape from the fittings before tightening. The specific procedures for connecting the gauges to a system containing refrigerant are discussed below.

On systems where it is certain that both pressures are above 0 psig (100.68 kPa), use these procedures:

1. Front-seat the valves on the gauge manifold.
2. Back-seat the system service valves. This is to isolate the gauge ports from the rest of the system.
3. Make the hose connections to the system. Loosen one end of the center hose.
4. Crack the system service valves off the back seat. Do this slowly to prevent a sudden inrush of high-pressure gas to the gauge.
5. Crack open one valve on the gauge manifold and allow a small amount of refrigerant vapor to escape out the center hose for a few seconds. Close the gauge manifold valve and repeat this process with the other valve.
6. The gauge manifold is now connected to the system and is ready for use (see Fig. 4-4).

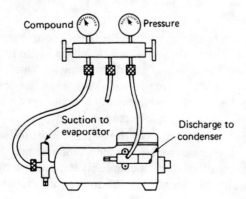

Figure 4-4. Gauge manifold connected to system.

On systems where the low-side pressure is below 0 psig (100.68 kPa), use these procedures:

1. Front-seat the hand valves on the gauge manifold.
2. Back-seat the system service valves. This is to isolate the gauge ports from the rest of the system.
3. Make the hose connections to the system. Tighten the hose connection to the discharge valve, loosen the hose connection to the suction service valve, and plug the center hose connection to prevent the escape of refrigerant at this point.
4. Crack the system discharge service valve off the back-seat. Do this slowly to prevent a sudden inrush of high-pressure gas to the gauge.
5. Crack open the high-side gauge manifold valve and allow a small amount of refrigerant to escape out the low-side hose connection at the system service valve. Tighten the hose connection after a few seconds. Close the gauge manifold hand valve.
6. Crack the system suction service valve off the back seat.
7. The gauge manifold is now connected to the system and is ready for use (see Fig. 4-4).

Gauges are delicate instruments and should be treated with care. Do not drop the gauges or subject them to pressures higher than the maximum pressure shown on the scale. Gauges should be kept in adjustment so that proper pressures are indicated.

4-3 PROCEDURE FOR CHECKING A THERMOCOUPLE

1. Install an adapter between the thermocouple and power unit.
2. Light pilot. Follow the lighting instructions on the furnace.
3. Set the meter on the millivoltage scale.
4. To check the open-circuit voltage, remove the thermocouple from the adapter and touch one meter lead to the outer sheath of the thermocouple, and the other meter lead to the inner wire of the thermocouple. This is dc voltage. If a negative reading is indicated, change the meter leads. Be sure to keep the pilot burning during this test. The meter should indicate approximately 30 mV. If less than 25 mV is indicated, replace the thermocouple (see Fig. 4-5).
5. To check closed-circuit voltage, connect the thermocouple to the adapter with the pilot burning. Touch one meter lead to the adapter terminal and the other lead to the outer sheath of the thermocouple. If the reading indicated is less than 17 mV, replace the thermocouple or clean the pilot burner, or both (see Fig. 4-6).

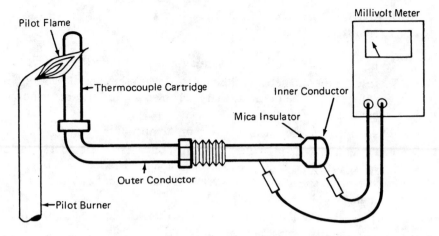

Figure 4-5. Open circuit thermocouple test.

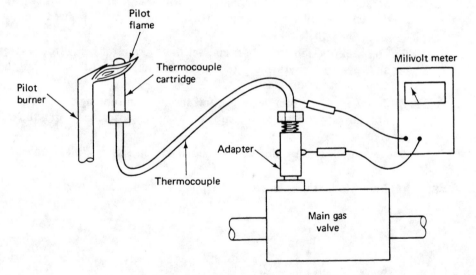

Figure 4-6. Closed circuit thermocouple test.

6. To check the thermocouple dropout voltage, connect the meter as outlined in step 5. Turn off the pilot and observe the meter. When a dull thud in the gas valve is heard, the dropout voltage is indicated on the meter. The dropout voltage should be about 8 mV. If the dropout is above 8 mV, the pilot stat power unit is defective and needs replacement. There is no correction for a dropout voltage below this.

7. Remove the adapter, reconnect the thermocouple to the power unit, light the pilot, and replace covers and panels.

4-4 PROCEDURE FOR ADJUSTING FAN
AND LIMIT CONTROL

1. Light the pilot. Follow the lighting instructions on the furnace.

2. Insert a thermometer as close as possible to the control sensing element. This will generally be in the discharge air plenum directly above the fan and limit control (see Fig. 4–7).

3. Turn the thermostat up above room temperature.

4. Observe the thermometer. The fan should start at about 135 to 150°F (57 to 66°C). If not, adjust the control to compensate for the difference. *Example:* If the fan starts at 120°F (49°C), adjust the fan "on" setting 15°F (8.34°C) higher.

5. To check the limit control, restrict the flow of air through the furnace by blocking the return air opening.

6. Observe the thermometer. The main burner flame should go out at a maximum of 200°F (93°C). If the flame goes out between 180°F (82°C) and 200°F (93°C), no adjustment is required. If adjustment is required, follow the example in step 4.

7. To check the fan "off" setting, set the thermostat below room temperature and remove the return air restriction used in step 6. Allow the fan to run and cool down the furnace.

8. Observe the thermometer. The fan should stop running at 100°F (37.8°C). If not, adjust the control to compensate for the difference, using the example in step 4.

9. Repeat all steps outlined above and check the settings. Make any adjustments required.

10. Remove the thermometer and replace all panels and covers.

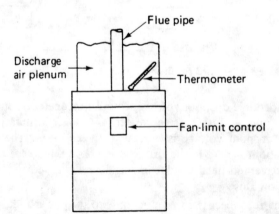

Figure 4-7. Checking fan and limit control operation.

4-5 PROCEDURE FOR PURGING NONCONDENSABLES FROM THE SYSTEM

Air is considered a noncondensable under the pressures and temperatures normally encountered in an air-conditioning or refrigeration system. Air can enter the refrigerant circuit in several ways. The most common ways are a leak in the low side of the system or improper connection and use of the service gauges. In some cases it may not be practical or desirable to purge and complete refrigerant charge and evacuate the system. However, the air must be removed to prevent damage to the system due to chemical reactions and to help keep the system operating efficiently.

The air will normally be trapped in the top of the receiver and the condenser because of the liquid seal at the receiver or condenser outlet. Air in the system may be detected by a higher-than-normal condensing pressure caused by the trapped air.

To purge noncondensables from the system, use the following procedure:

1. Locate and remove the source of noncondensables.
2. Connect the gauges to the system (see Fig. 4-4).
3. If possible, pump down the system.
4. Stop the unit. Leave the condenser fan running on air-cooled units or block the water valve open on water-cooled units and leave the pump running. Allow the unit to cool down for approximately 10 minutes. During this time the noncondensables will rise to the top of the condenser.
5. If purge valves are on the unit, use them for the purging process. If not, the gauge port on the compressor discharge service valve may be used. To purge, slowly open the purge valve. Allow the vapor to bleed off very slowly for only a short period of time. Close the valve and let the unit sit idle. Be sure to purge slowly for only short periods to prevent the boiling off of excess refrigerant and the remixing of the air and refrigerant. After the system has set idle for a few minutes, repeat the purging process. Repeat the process three or four times.
6. Start the system and check the discharge pressure after a few minutes of operation. If the discharge pressure is still abnormally high, repeat the purging process starting with step 4. Repeat steps 4, 5, and 6 until satisfactory operation is obtained.
7. Put the system back in normal operating condition.

4-6 PROCEDURE FOR PUMPING DOWN A SYSTEM

Pumping down a system is the process generally used to put all the refrigerant into the condenser or receiver. Pumping down is used to save the refrigerant when work is required on components in the low side of the refrigerant system. Pumping down a system can be accomplished on systems equipped with service valves and a liquid line shutoff valve.

To pump down a system, use the following procedure:

1. Install the gauge manifold on the system.
2. Start the unit.
3. Front-seat the liquid line (king) valve.
4. Observe both the suction and discharge pressures. If there is a sharp increase in the discharge pressure, stop the compressor. Check to determine the reason for the increase in discharge pressure. If the receiver and condenser are full of refrigerant, the remaining charge must be removed from the system.

   When the suction pressure is reduced to about 1 or 2 psig (107.57 or 114.46 kPa), stop the compressor and observe the gauges. If the suction pressure increases to 10 or 15 psig (169.58 or 204.03 kPa), start the compressor and pump down the system to 1 or 2 psig again and stop the compressor. The suction pressure should remain at around this pressure. If not, repeat the pump-down process. If the pump-down process is repeated more than three times, the compressor discharge valves may be leaking. In this case the discharge service valve must be closed to prevent refrigerant bleeding into the low side of the system. *Caution:* Do not start the compressor when the discharge valve is closed.

5. On units equipped with a low-pressure control set at a pressure higher than 1 or 2 psig, it will be necessary to electrically bypass the control to keep the compressor running while pumping down the system.
6. Relieve any remaining pressure on the low side of the system by opening the low-side hand valve on the gauge manifold. Do not attempt to weld or solder on a system having refrigerant pressure inside.
7. The necessary repairs can now be made. It is desirable to install a new liquid-line drier before recharging the system.
8. To put the system back in operation, open the compressor discharge service valve and open the liquid line (king) valve. Allow a small amount of refrigerant to escape through the gauge manifold; then close the gauge manifold low-side hand valve.
9. Start the unit and check the refrigerant charge. Add any required refrigerant to bring the system to full charge.

4-7 PROCEDURE FOR PUMPING OUT A SYSTEM

Pumping out a system is the process used to save the refrigerant in a system that does not have service valves, when repairs are to be made. This procedure is also used when repairs are to be made to the high side of the system and a system pump-down cannot be done. To accomplish this process, a portable condensing unit and a clean, dry refrigerant cylinder are required. The portable condensing unit is used to pump the refrigerant from the system and discharge it into the refrigerant cylinder, where it is stored while the repairs are being made. The refrigerant is then charged back into the system.

To pump out a system, use the following procedure:

1. Stop the unit.
2. Connect the gauge manifold to the system (see Fig. 4–4).
3. Connect the center hose on the gauge manifold to the suction service valve on the portable condensing unit.
4. Connect the liquid-line connection on the portable condensing unit to the valve on the refrigerant cylinder. Leave this connection loose for purging air from the lines and from the portable condensing unit (see Fig. 4–8).
5. Slowly crack the system service valves off the back seat until pressure is indicated on the gauges. Open the service valves on the portable condensing unit.
6. Open the hand valves on the gauge manifold and allow the refrigerant to blow out the connection on the refrigerant cylinder for a few seconds. Tighten the hose connection on the refrigerant cylinder.
7. Open the refrigerant cylinder valve.
8. Start the portable condensing unit and pump the refrigerant from the system and into the cylinder. To prevent overloading of the portable condensing unit, regulate the suction pressure by partially closing the hand valves on the gauge manifold. *Caution:* Be sure not to overcharge a cylinder with refrigerant. Use as many cylinders as required by weight.
9. Remove the refrigerant until only 1 or 2 psig pressure (107.57 or 114.46 kPa) is left inside the system. Close off all valves to prevent the refrigerant escaping from the cylinder and portable condensing unit.

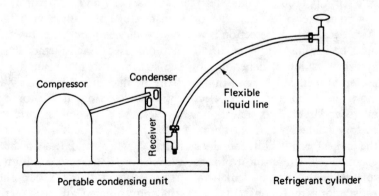

Figure 4-8. Connections for portable condensing unit to refrigerant cylinder.

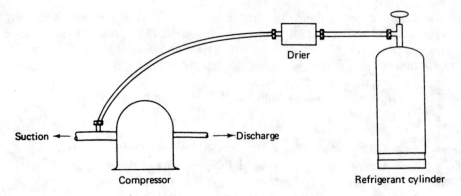

Figure 4-9. Liquid drier in charging line.

10. Purge any remaining pressure from the refrigeration system and make the required repairs. It is desirable to install a new liquid line drier before charging the system with refrigerant.

11. To put the system back in operation, a complete evacuation should be completed. Then charge the refrigerant from the cylinder and the portable condensing unit back into the system. It may be desirable to install a drier in the charging line to remove any contaminants from the cylinder or portable condensing unit (see Fig. 4-9).

4-8 PROCEDURE FOR LEAK TESTING

When the refrigerant has escaped from the system, the leak must be found and repaired or the refrigerant will escape again. Refrigeration systems must be gastight, for two reasons: (1) any leakage will result in a loss of refrigerant charge, and (2) leaks will allow air and moisture to enter the system when the pressure is reduced below 0 psig (100.68 kPa). Refrigerant leaks can and do occur at any time and at any point due to low-quality workmanship, age of equipment, and vibration.

Because leak detection is a common service procedure, the service technician must be familiar with the various methods used. Each method has its advantages and disadvantages, depending on the circumstances surrounding the system. All preferred methods require that pressure be introduced into the system. The three most popular methods are: electronic leak detection, halide-torch leak detection, and soap-bubble leak detection. However, the electronic method is not very satisfactory in a heavy concentration of refrigerant because it is extremely sensitive; and the halide torch is not generally satisfactory when small leaks or blowing winds are encountered. Under most other conditions these two tests are found most satisfactory. Because electronic leak detectors are very sensitive, the soap-bubble test is most satisfactory for finding small leaks in nonventilated areas. Sometimes more than one of these methods may be used to locate one leak.

Evacuation is not a recommended leak detection method, for many reasons. First, evacuation will not pinpoint a leak. Second, a flake of paint or a grain of sand may cover the leak during evacuation, preventing the loss of vacuum in the system. Third, more air and moisture are drawn into the system, requiring more evacuation for their removal. Therefore, one of the three methods noted above is preferred.

To leak-test a system, use the following procedure:

1. Stop the unit. This is to reduce air movement as much as possible.
2. Choose the leak detection method best suitable for the situation encountered.
3. Connect the gauge manifold to the refrigerant system.
4. Be sure that the system has at least 35 psig (341.83 kPa) inside it. If not, increase the pressure by adding refrigerant vapor to the system. Be sure to use the same type of refrigerant as that in the system.
5. Test each joint, fitting, and gasket in the entire system. Mark each leak as it is found. The repairs can be made after all leaks have been located. Any joints or areas that have signs of oil residue should be given special attention. If no leak is found with the halide torch or the electronic leak detector, use a soap-bubble solution to be sure that no leaks exist.
6. Either pump down the system, pump out the system, or purge the refrigerant from the system, depending on the amount of refrigerant in the system. Usually, a refrigerant charge of less than 10 lb (4.5359 kg) is purged. A charge of more than 10 lb (4.5359 kg) is generally saved.
7. Repair the leaks and leak-test to be sure that all leaks have been stopped.
8. Evacuate the system.
9. Install new liquid-line driers on the system.
10. Recharge the system with the proper amount of clean, dry refrigerant.

4-9 PROCEDURE FOR EVACUATING A SYSTEM

Evacuation is the process used to remove air and moisture from a refrigeration system. Evacuation is accomplished by use of pumps specially designed for this purpose. A discarded refrigeration compressor is not suitable. Never run a compressor while the system is evacuated. To do so may result in serious damage to the motor winding. Evacuation is required any time a system has become contaminated or the compressor or system has been exposed to the atmosphere for long periods.

Purging a system will remove a good portion of the air, and driers will remove a good part of the moisture from the system—but only up to the capacity of the drier. Therefore, there are still contaminants left in the system, and evacuation is the best means of being reasonably sure that the system is free of these contaminants.

There are basically two evacuation procedures: (1) simple evacuation, and (2) triple evacuation. Simple evacuation is used on systems containing only a minimum amount of contaminants. Triple evacuation is used on systems containing a greater amount of contaminants.

To use the simple evacuation method, use the following procedure:

1. Connect the gauge manifold to the system (see Fig. 4–4).
2. Purge all pressure from the system by opening the system service valves and the gauge manifold hand valves.
3. Connect the center hose on the gauge manifold to the vacuum pump (see Fig. 4–10).
4. Start the vacuum pump and pump a vacuum of at least 1000 microns. A vacuum of 500 microns is preferable.
5. Close off all gauge manifold hand valves.
6. Stop the vacuum pump. Do not stop the vacuum pump before closing the gauge manifold hand valves. This is to prevent air entering the system.
7. Disconnect the center hose of the gauge manifold from the vacuum pump and connect it to a cylinder containing the proper refrigerant (see Fig. 4–11).
8. Open the cylinder valve.

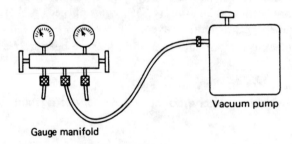

Vacuum pump

Gauge manifold

Figure 4–10.   Gauge manifold connected to vacuum pump.

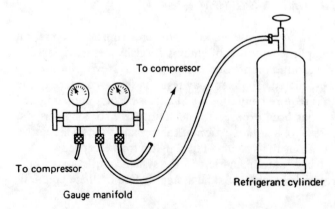

To compressor

To compressor

Gauge manifold

Refrigerant cylinder

Figure 4–11.   Gauge manifold connected to refrigerant cylinder.

9. Loosen the center hose connection at the gauge manifold. Purge the hose for a few seconds; then tighten the connection.
10. Open the gauge manifold hand valves and admit refrigerant into the system.
11. Close the high-side hand valve on the gauge manifold.
12. Start the unit and add the proper charge of refrigerant.

To use the triple-evacuation method, use the following procedure:

1. Connect the gauge manifold to the system (see Fig. 4-4).
2. Purge all pressure from the refrigerant system by opening the systems service valves and the gauge manifold hand valves.
3. Connect the center hose on the gauge manifold to the vacuum pump (see Fig. 4-11).
4. Start the vacuum pump and pump a vacuum of approximately 1500 microns.
5. Close off the gauge manifold hand valves.
6. Stop the vacuum pump. Do not stop the vacuum pump before closing the gauge manifold hand valves. This is to prevent air from entering the system.
7. Disconnect the center hose of the gauge manifold from the vacuum pump and connect it to a cylinder containing the proper refrigerant.
8. Open the cylinder valve.
9. Loosen the center hose connection at the gauge manifold. Purge the hose for a few seconds; then tighten the connection.
10. Open the gauge manifold hand valves and admit refrigerant into the system until a pressure of about 5 psig (135.13 kPa) is indicated on the gauges.
11. Close the refrigerant cylinder valve and the gauge manifold hand valves.
12. Disconnect the hose from the cylinder.
13. Open the gauge manifold hand valves and purge the pressure from the system.
14. Repeat steps 3 through 13.
15. Repeat steps 3 through 9 only. Pump a vacuum of 500 microns rather than 1500 microns.
16. Open the gauge manifold hand valves and admit refrigerant into the system until cylinder pressure is indicated on the gauges.
17. Close the high-side gauge manifold hand valve.
18. Start the unit and add the proper charge of refrigerant to the system.

4-10 PROCEDURE FOR CHARGING REFRIGERANT INTO A SYSTEM

The performance of a refrigeration and air-conditioning system is highly dependent on the proper charge of refrigerant in the system. A system that is undercharged will operate with a starved evaporator. Low suction pressures, loss of system capacity, and possible compressor overheating are the results of an undercharged system. On the other hand, an overcharge of refrigerant will back up in the condenser. High discharge pressures and liquid refrigerant flooding the compressor with possible compressor damage are the results of an overcharged system. Larger systems can tolerate a reasonable amount of overcharging or undercharging without severe effects, but some of the smaller systems have a critical charge. In critical charged systems they must be properly charged to obtain proper operation.

The amount of charge will depend on the size of the system in Btu, the length of the lines, the type of refrigerant, and the operating temperature. Therefore, each system must be considered separately. The unit nameplate will usually indicate what type of refrigerant is required and the approximate weight of the refrigerant required.

There are two methods of charging refrigerant into a system: (1) liquid charging, and (2) vapor charging. Liquid charging is much faster than vapor charging and is therefore used extensively on large field built-up systems. Never charge liquid refrigerant into the compressor suction or discharge service valves. Liquid entering the compressor can damage the compressor valves. Vapor charging is normally done only when small amounts of refrigerant are added to a system. Vapor charging also allows the refrigerant to be charged into the compressor suction service valve.

To use the liquid charging method, use the following procedure:

1. Connect the gauge manifold to the system (see Fig. 4–4).
2. Open the system service valves and purge air from the lines.
3. Connect a charging line to the liquid-line (king) valve. This line should include a liquid drier to prevent contaminants from entering the system.
4. Connect the charging line to the liquid valve on a cylinder of the proper type of refrigerant (see Fig. 4–12).
5. Open the refrigerant cylinder liquid valve.
6. Loosen the charging line connection at the liquid-line (king) valve and allow the refrigerant to escape for a few seconds.

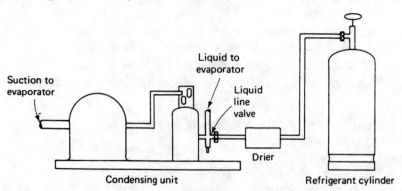

Figure 4–12. Connections for liquid charging a system.

7. If the correct charge weight is known, the refrigerant cylinder can be weighed to determine when sufficient refrigerant has been charged into the system.
8. Close the liquid-line (king) valve and start the compressor. Allow the liquid refrigerant to enter the liquid line until the correct weight has been charged into the system (see step 10).
9. If the correct charge weight is not known, the liquid-line valve must be opened periodically and the system operation observed. If more refrigerant is needed, close the liquid-line valve again and charge more refrigerant into the system. Repeat this process until the proper charge is indicated (see step 10).
10. Closely watch the discharge pressure gauge. A sudden increase in discharge pressure indicates that the capacity of the condenser and receiver has been reached. Stop charging the unit immediately and open the liquid-line valve. Any additional refrigerant must be charged into the system by the vapor method.

To use the vapor charging method, use the following procedure:

1. Connect the gauge manifold to the system (see Fig. 4–4).
2. Connect the center line on the gauge manifold to a cylinder of the proper refrigerant (see Fig. 4–11).
3. Open the refrigerant cylinder valve and the hand valves on the gauge manifold.
4. Loosen the hose connections at the system service valves and allow refrigerant to escape for a few seconds. Retighten the connections.
5. Close the gauge manifold hand valves.
6. Crack the system service valves.
7. Start the unit.
8. Open the low-side gauge manifold hand valve and charge refrigerant into the system until the proper amount has been charged into the system.
9. Closely watch the discharge pressure gauge during the charging process to be certain that the system is not overcharged.

4-11   PROCEDURE FOR DETERMINING THE PROPER REFRIGERANT CHARGE

Determining the proper refrigerant charge is an important procedure. A system that does not contain the proper charge of refrigerant will not operate to maximum efficiency. There are several ways to determine if a system is properly charged, such as weighing the charge, using a sight glass, using a liquid-level indicator, using the liquid subcooling method, using the superheat method, and using the manufacturer's charging charts.

To use the charge weight method, use the following procedure:

1. Connect the gauge manifold to the system service valves (see Fig. 4–4).
2. Purge all pressure from the system. Open both hand valves on the gauge manifold.
3. Connect the centerline from the gauge manifold to a vacuum pump (see Fig. 4–10).
4. Start the vacuum pump and pump a deep vacuum on the system.
5. Close the gauge manifold hand valves.
6. Stop the vacuum pump. Do not stop the vacuum pump before closing the gauge manifold hand valves. This is to prevent air from entering the system.
7. Disconnect the center hose from the vacuum pump and connect it to a cylinder of the proper type of refrigerant (see Fig. 4–11). This cylinder may be a charging cylinder loaded with the proper amount of refrigerant or a large cylinder placed on an accurate scale. Small systems require a charging cylinder, whereas large systems require a large cylinder.
8. Open the cylinder valve.
9. Loosen the center hose connection at the gauge manifold. Purge the hose for a few seconds; then tighten the connection.
10. Open both hand valves on the gauge manifold and charge the refrigerant into the system. This must be done suddenly so that all the refrigerant will enter the system before the vacuum vanishes.
11. Start the unit and check the operation.

To use the sight glass method, use the following procedure:

1. Start the unit and allow it to operate for several minutes.
2. Check the flow of refrigerant through the sight glass. A flashlight may be needed to adequately see the flow (see Fig. 4–13).

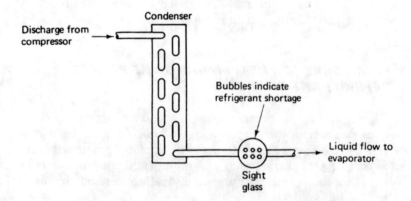

Figure 4-13.   Checking refrigerant charge using a sight glass.

3. A steady stream of bubbles indicates that the system is low on refrigerant. If these bubbles are intermittent, allow the system to operate awhile longer to see if they will disappear. If the bubbles remain, the system is low on refrigerant.

4. Connect the gauge manifold to the system service valves (see Fig. 4–4).

5. Connect the center line to a cylinder of the proper type refrigerant (see Fig. 4–11).

6. Crack the system service valves and open the gauge manifold hand valves and loosen the connection on the refrigerant cylinder. Purge the lines for a few seconds. Tighten the connection.

7. Close the hand valves on the gauge manifold.

8. Open the refrigerant cylinder valve.

9. Open the low-side hand valve on the gauge manifold and admit refrigerant into the system while observing the sight glass and the discharge pressure gauge. When the bubbles disappear, close the gauge manifold low-side hand valve. Observe the sight glass. If the bubbles reappear, add more refrigerant into the system. Repeat this process until the bubbles do not reappear. A sudden increase in the discharge pressure indicates a system overcharge. Stop charging the unit and remove some of the refrigerant.

To use the liquid-level indicator method, use the following procedure:

1. Start the unit and allow it to operate 10 to 15 minutes.

2. After the system has been in operation long enough for the pressures to stabilize, crack the liquid-level test port. A continuous flow of liquid refrigerant from the port indicates sufficient charge. A continuous flow of vapor from the port indicates a shortage of refrigerant. The test port is usually located in the lower section of the condenser. Some units may have a liquid level indicator in the receiver tank (see Fig. 4–14).

3. To add refrigerant to the system, connect the gauge manifold to the system service valves. Close the gauge manifold hand valves. Connect the center

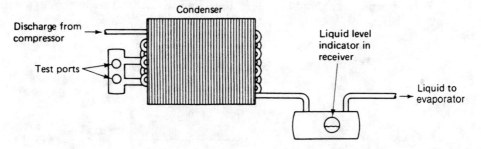

Figure 4-14.   Liquid level indicator parts.

charging line to a cylinder containing refrigerant of the proper type. Leave
this connection loose (see Fig. 4–11).

4. Crack the system service valves open until pressure is indicated on the gauges.
5. Open the gauge manifold hand valves. Allow refrigerant to escape from the
   hose connection on the cylinder for a few seconds; then tighten the connection.
6. Close the high-side gauge manifold hand valve.
7. Open the refrigerant cylinder valve and charge refrigerant into the system.
8. Periodically open the liquid-level test port and check for liquid refrigerant.
9. Continue to add refrigerant until a continuous stream of liquid refrigerant is
   indicated.
10. When liquid is indicated, close the low-side gauge manifold hand valve. Allow
    the unit to operate a few minutes and check for liquid again. If a continuous
    stream of liquid is not indicated, add more refrigerant and check again. Con-
    tinue this process until the proper charge is indicated.

To use the liquid subcooling method, use the following procedure:

1. Start the unit and allow it to operate for 10 to 15 minutes.
2. Strap a thermometer to the liquid line at the condenser outlet (see Fig. 4–15).
3. After the system has operated long enough for the pressures to stabilize, check
   the liquid temperature leaving the condenser as indicated on the thermometer
   installed in step 2.
4. Connect the gauge manifold to the system service valves. Close the gauge manifold hand valves (see Fig. 4–4).
5. Open the system service valves until pressure is indicated on the gauges.
6. Loosen the hose connections on the gauge manifold and allow the refrigerant
   vapor to escape for a few seconds; then retighten the connections.
7. Compare the liquid-line temperature to the condensing temperature. The condensing temperature is determined by relating the discharge pressure to the
   corresponding temperature on a pressure–temperature chart. The liquid-line
   temperature is indicated on the thermometer installed in step 2. The liquid line
   temperature should be approximately 5°F (2.78°C) below the condensing tem-

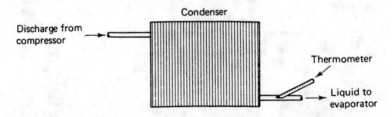

Figure 4–15. Checking liquid subcooling during the charging process.

perature. If less than 5°F (2.78°C), more refrigerant is needed in the system (see Table 4–1).

8. To add refrigerant, connect the center charging hose on the gauge manifold to the valve on a cylinder containing the proper type of refrigerant. Do not tighten this connection (see Fig. 4–11).

9. Open the hand valves on the gauge manifold. Allow the refrigerant to escape a few seconds; then tighten the connection of the cylinder valve.

10. Close the high-side hand valve on the gauge manifold.

11. Open the cylinder valve and charge refrigerant into the system.

12. Observe the thermometer on the liquid line and the discharge temperature. When the desired subcooling is obtained, close the low-side hand valve on the gauge manifold to stop adding refrigerant to the system.

13. Allow the system to operate a few minutes to allow the pressures to stabilize, and check the amount of subcooling. If additional subcooling is required, add more refrigerant to the system. Continue this process until the 5°F (2.78°C) subcooling remains stable.

To use the superheat method, use the following procedure:

1. Start the unit and allow it to operate for 10 to 15 minutes.

2. Strap a thermometer on the suction line about 6 in. from the compressor. Insulate the thermometer bulb so that accurate readings are indicated (see Fig. 4–16).

3. If a low-side pressure port is available, connect the low-side gauge to the system by connecting the low-side hose on the gauge manifold to the pressure port. If a pressure port is not available, strap a thermometer to a return bend about midway on the evaporating coil. Do not put the thermometer on a fin. Insulate the thermometer bulb so that accurate readings may be indicated (see Fig. 4–17).

4. If a pressure port is available, determine the difference between the suction-line temperature and the saturation temperature equivalent to the suction pres-

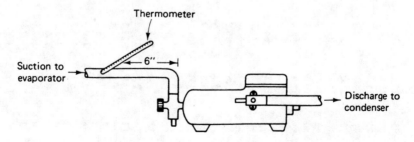

Figure 4-16.   Thermometer strapped to suction line.

TABLE 4-1 PRESSURE–TEMPERATURE CHART[a,b]

°F	R-12	R-13	R-22	R-500	R-502	R-717 Ammonia
-100	27.0	7.5	25.0	26.4	23.3	27.4
-95	26.4	10.9	24.1	26.7	22.1	26.3
-90	25.8	14.2	23.0	24.9	20.7	26.1
-85	25.0	18.2	21.7	19.0	19.0	25.3
-80	24.1	22.3	20.2	22.0	17.1	24.3
-75	23.0	27.1	18.5	21.7	15.0	23.2
-70	21.0	32.0	16.6	20.3	12.6	21.9
-65	20.5	37.7	14.4	18.8	10.5	20.4
-60	19.0	43.5	12.0	17.0	7.0	18.6
-55	17.3	50.0	9.2	15.0	3.6	16.6
-50	15.4	57.0	6.2	12.5	0.0	14.3
-45	13.3	64.6	2.7	10.4	2.1	11.7
-40	11.0	72.7	0.5	7.6	4.3	8.7
-35	8.4	81.5	2.6	4.6	6.7	5.4
-30	5.5	90.9	4.9	1.2	9.4	1.6
-28	4.3	94.9	5.9	0.2	10.5	0.0
-26	3.0	98.9	6.9	0.9	11.7	0.8
-24	1.6	103.0	7.9	1.6	13.0	1.7
-22	0.3	107.3	9.0	2.4	14.2	2.6
-20	0.6	111.7	10.2	3.2	15.5	3.6
-18	1.3	116.2	11.3	4.1	16.9	4.6
-16	2.1	120.8	12.5	5.0	18.3	5.6
-14	2.8	125.7	13.8	5.8	19.7	6.7
-12	3.7	130.5	15.1	6.8	21.2	7.9
-10	4.5	135.4	16.5	7.8	22.8	9.0
-8	5.4	140.5	17.9	8.8	24.4	10.3
-6	6.3	145.7	19.3	9.9	26.0	11.6
-4	7.2	151.1	20.8	11.0	27.7	12.9
-2	8.2	156.5	22.4	12.1	29.4	14.3
0	9.2	162.1	24.0	13.3	31.2	15.7
2	10.2	167.9	25.6	14.5	33.1	17.2
4	11.2	173.7	27.3	15.7	36.0	18.8
6	12.3	179.8	29.1	17.0	37.0	20.4
8	13.5	185.9	30.9	18.4	39.0	22.1
10	14.6	192.1	32.8	19.7	41.1	23.8
12	15.8	198.6	34.7	21.2	43.2	25.6
14	17.1	205.2	36.7	22.6	45.5	27.5

°F	R-12	R-13	R-22	R-500	R-502	R-717 Ammonia
16	18.4	211.9	38.7	24.1	47.7	29.4
18	19.7	218.8	40.9	25.7	50.1	31.4
20	21.0	225.7	43.0	27.3	52.5	33.5
22	22.4	233.0	45.3	28.9	54.9	36.7
24	23.9	240.3	47.6	30.6	57.4	37.9
26	25.4	247.8	49.9	32.4	60.0	40.2
28	26.9	255.5	52.4	34.2	62.7	42.6
30	28.5	263.2	54.9	36.0	65.4	45.0
32	30.1	271.3	57.5	37.9	68.2	47.6
34	31.7	279.5	60.1	39.9	71.7	50.2
36	33.4	287.8	62.8	41.9	74.1	52.9
38	36.2	296.3	65.6	43.9	77.1	55.7
40	37.0	304.9	68.5	46.1	80.2	58.6
45	41.7	327.5	76.0	51.6	86.3	66.3
50	46.7	351.2	84.0	57.6	96.9	74.5
55	52.0	376.1	92.6	63.9	106.0	83.4
60	57.7	402.3	101.6	70.6	115.6	92.9
65	63.8	429.8	111.2	77.8	125.8	103.1
70	70.2	458.7	121.4	85.4	136.6	114.1
75	77.0	489.0	132.2	93.5	148.0	125.8
80	84.2	520.8	143.6	102.0	159.9	138.3
85	91.8	—	155.7	111.0	172.5	151.7
90	99.8	—	168.4	120.6	185.8	165.9
95	108.3	—	181.8	130.6	199.7	181.1
100	117.2	—	195.9	141.2	214.4	197.2
105	126.6	—	210.8	152.4	229.7	214.2
110	136.4	—	226.4	164.1	245.8	232.3
115	146.8	—	242.7	176.5	262.6	251.5
120	157.7	—	259.9	189.4	280.3	271.7
125	169.1	—	277.9	203.0	298.7	293.1
130	181.0	—	296.8	217.2	318.0	315.0
135	193.5	—	316.6	232.1	338.1	335.0
140	206.6	—	337.3	247.7	369.1	365.0
145	220.3	—	368.9	266.1	381.1	390.0
150	234.6	—	381.5	261.1	403.9	420.0
155	249.9	—	405.1	296.9	427.8	450.0
160	265.1	—	429.8	317.4	452.6	490.0

[a]Bold Figures = inches mercury vacuum; light figures = psig.

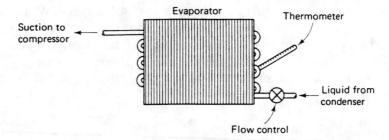

Figure 4-17. Thermometer strapped to evaporating coil return bend.

sure with the unit running. The saturation temperature is determined by relating the suction pressure to the corresponding temperature on a pressure-temperature chart. The temperature difference should be approximately 20 to 30°F (11.11 to 16.67°C) (see Table 4-1).

   If no pressure port is available, check the difference between the two thermometers. The difference in temperature should be approximately 20 to 30°F (11.11 to 16.67°C). When the unit is operating at a normal operating condition, a superheat of 20 to 30°F (11.11 to 16.67°C) is satisfactory. A superheat lower than 20°F (11.11°C) indicates an overcharge of refrigerant. A superheat higher than 30°F (11.11°C) indicates an undercharge of refrigerant.

5. To add refrigerant, connect the low-side gauge on the gauge manifold to the pressure port on the system. If a low-side pressure port is not available, a saddle valve must be installed (see Fig. 4-18).

6. Connect the center charging hose on the gauge manifold to the valve on a cylinder of the proper type refrigerant. Do not tighten this connection (see Fig. 4-11).

7. Open the gauge manifold hand valve and allow the refrigerant to escape for a few seconds. Tighten the connection on the cylinder valve.

8. Open the refrigerant cylinder valve and charge refrigerant into the system.

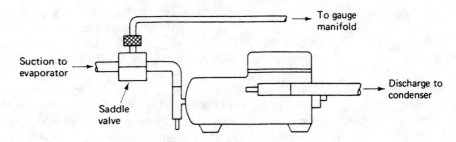

Figure 4-18. Installation of saddle valve.

9. Observe the superheat by the method used in step 4. When the proper superheat is reached, stop charging refrigerant into the system.

10. Allow the unit to continue running until the pressures and temperatures have stabilized. If more refrigerant is needed, repeat the charging process until the desired superheat is stabilized.

To use the manufacturer's charging chart method, use the following procedure:

1. Start the unit and allow it to operate for 10 to 15 minutes.
2. Connect the gauge manifold to the system service valves.
3. Connect the center charging hose on the gauge manifold to the valve on a cylinder of the proper type of refrigerant. Do not tighten this connection.
4. Open the hand valves on the gauge manifold and allow refrigerant to escape from the loose connection for a few seconds. Tighten the connection on the cylinder valve.
5. Close the high-side hand valve on the gauge manifold.
6. Obtain a copy of the manufacturer's charging chart for that model unit.
7. Compare the pressures to those indicated on the chart.
8. If more refrigerant is needed, open the valve on the cylinder and add refrigerant to the system.
9. Stop charging after a few minutes by closing the low-side hand valve on the gauge manifold and compare the pressure readings indicated on the chart. Repeat this procedure until the system pressures compare with those indicated on the chart.

4-12 PROCEDURE FOR DETERMINING THE COMPRESSOR OIL LEVEL

All refrigeration compressors require a specific amount of oil. This oil is required for lubrication of the moving parts and helps to make a refrigerant seal between the components. An abnormally low oil level will probably result in a loss of lubrication and compressor damage. An excess of lubricating oil will result in oil slugging, probable damage to the compressor valves, and lost system efficiency due to oil logging of the evaporating coil.

To use the sight glass method, use the following procedure:

1. Start the unit and allow it to operate for 10 to 15 minutes.
2. Check the oil level in the sight glass (see Fig. 4-19). A flashlight may be needed to determine the level accurately. The oil level should be at or slightly above the center of the sight glass while the unit is operating. If less than this, oil should be added. If more than this, the excess oil should be removed.

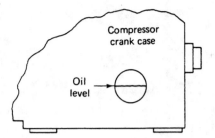

Figure 4-19. Checking compressor oil level.

On sealed systems that do not have an oil sight glass, determining the amount of oil in the compressor is difficult. When a leak occurs and the amount of oil lost is small and can be reasonably calculated, add that amount to the system. If a large amount of oil has been lost, the compressor should be removed, the oil drained, and the correct amount added to the compressor before placing it back in operation.

4-13 PROCEDURE FOR ADDING OIL TO A COMPRESSOR

Adding oil to a compressor is often required, and the service technician should be familiar with the various procedures used. There are three procedures used, depending on the type of system encountered and the type of tools at hand: (1) the open system method, (2) the closed system method, and (3) the oil pump method.

To use the open system method, use the following procedure:

1. Connect the gauge manifold to the system service valves (see Fig. 4-4).
2. Close the gauge manifold hand valves and open the compressor service valves.
3. Start the unit.
4. Front-seat the compressor suction service valve and run the unit until the low-side pressure is reduced to 1 or 2 psig (107.57 or 114.46 kPa). The low-pressure switch may need to be bypassed.
5. Stop the compressor.
6. Front-seat the compressor discharge service valve.
7. Open the low-side hand valve on the gauge manifold and relieve the slight pressure remaining in the system.
8. Remove the oil fill plug and pour the oil into the compressor crankcase until the proper level is reached. Use caution to prevent contamination of the oil (see Fig. 4-20).
9. Close the low-side hand valve on the gauge manifold.
10. Slightly open the compressor suction service valve and allow a small amount of refrigerant to escape out the oil fill hole.
11. Close off the compressor suction service valve.

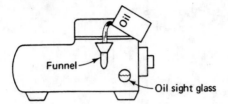

Figure 4-20. Pouring oil into a compressor.

12. Replace the oil fill plug.
13. Back-seat the compressor service valves.
14. Start the compressor and check the oil level in the compressor.
15. Remove the gauge manifold from the system if the proper level is present.

To use the closed system method, use the following procedure:

1. Connect the gauge manifold to the system service valves (see Fig. 4-4).
2. Place the center charging line in a container of clean dry oil (see Fig. 4-21).
3. Open the system service valves until a pressure is indicated on the gauges.
4. Slightly open the low-side hand valve on the gauge manifold and purge a small amount of refrigerant through the lines and the oil.
5. Front-seat the system service valves.
6. Start the unit and draw a vacuum on the compressor crankcase.
7. Open the low-side hand valve on the gauge manifold and draw the oil into the compressor through the service valve. Be sure that the center line remains in the oil to prevent air entering the system.
8. When a sufficient amount of oil has been drawn into the compressor, close the hand valve on the gauge manifold.
9. Back-seat the system service valves, and place the system in normal operation.

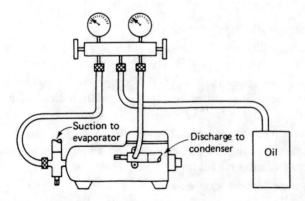

Figure 4-21. Adding oil to a closed system.

To use the oil pump method, use the following procedure:

1. Connect the low-side gauge of the gauge manifold to the system suction service valve.
2. Close the low-side hand valve on the gauge manifold.
3. Crack the system suction service valve until a pressure is indicated on the gauge.
4. Connect the center charging line to the oil pump. Do not tighten this connection.
5. Open the low-side gauge manifold hand valve and purge refrigerant through the loose connection for a few seconds; then tighten the connection.
6. Place the oil pump in a container of clean dry oil.
7. Open the gauge manifold hand valve completely.
8. Move the system service valve to the midway position.
9. Pump oil into the system until the proper level is reached (see Fig. 4-22).
10. Back-seat the system service valve.
11. Close the low-side hand valve on the gauge manifold.
12. Remove the gauge manifold and place the system in normal operation.

4-14 PROCEDURE FOR LOADING A CHARGING CYLINDER

Charging cylinders are ideal for charging systems that use less than 5 lb (2.2680 kg) of refrigerant. Charging cylinders are calibrated in ounces for each type of refrigerant with which they are designed to be used. Therefore, it is possible to charge the exact amount of refrigerant into the system. The service technician should be familiar with the steps involved in loading a charging cylinder.

To load a charging cylinder, use the following procedure:

1. Connect the low-pressure gauge of the gauge manifold to the charging cylinder (see Fig. 4-23).

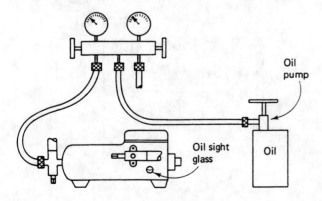

Oil
pump

Oil sight
glass

Oil

Figure 4-22. Adding oil to a system with an oil pump.

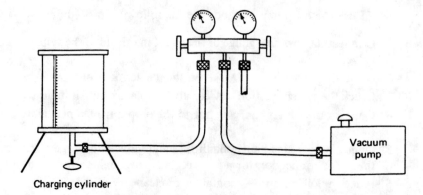

Figure 4-23.   Connections for evacuating a charging cylinder.

2. Open the hand valve on the charging cylinder.

3. Open the low-side hand valve on the gauge manifold and purge all pressure from the charging cylinder.

4. Connect the center charging line on the gauge manifold to the vacuum pump.

5. Start the vacuum pump and draw as deep a vacuum as possible with the pump.

6. Close the low-side hand valve on the gauge manifold.

7. Remove the center charging line from the vacuum pump and connect it to the valve on a cylinder of the proper-type refrigerant. If there is a liquid valve, connect the line to it (see Fig. 4-24).

8. Open the refrigerant cylinder valve and loosen the center charging line connection at the gauge manifold. Allow the refrigerant to escape for a few seconds, and then retighten the connection.

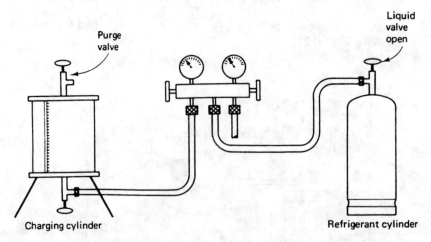

Figure 4-24.   Connections for loading a charging cylinder.

9. Invert the refrigerant cylinder so that liquid refrigerant will enter the charging line. If the line is connected to a liquid valve, the cylinder does not need to be inverted.

10. Open the low-side hand valve on the gauge manifold and allow liquid refrigerant to be drawn into the charging cylinder.

11. When the desired amount of refrigerant is drawn into the charging cylinder, close the refrigerant cylinder valve. If the refrigerant stops flowing before the desired amount is in the charging cylinder, the vent valve on top of the charging cylinder may be opened to permit the escape of vapor, thus allowing more liquid to enter (see Fig. 4-24).

12. Close the charging cylinder hand valve and disconnect the lines from the charging cylinder. Be sure that any liquid refrigerant in the lines does not come in contact with the skin or eyes.

4-15 *PROCEDURE FOR CHECKING COMPRESSOR ELECTRICAL SYSTEMS*

Potential Starting Relay System with Two-Terminal Overload

Potential (voltage) starting relays are nonpositional. The contacts are normally closed. These relays are generally used on single-phase compressors of ½ hp and larger. Use the following procedure to check this type of system (see Fig. 4-25):

1. Check to make certain that the proper voltage is being supplied to the unit. If proper voltage is supplied, continue to make the following checks. If the wrong voltage is being supplied, correct the problem with the power supply.

2. Disconnect the electrical power from the unit.

3. If a fan motor is used, disconnect one of the leads.

4. Make the following checks with an ohmmeter.

5. Check the continuity between points 1 and 2. Refer to Fig. 4-25. If no continuity is found, the control contacts are open. Close the control contacts by setting the control to demand operation. If continuity is found, continue to make the following checks. If no continuity is found with the control on demand, replace the control.

6. Check the continuity between points 3 and 4. If continuity is found, continue to make the following checks. If no continuity is found, wait about 10 minutes and check the continuity again. The overload may be tripped. If no continuity is found now, replace the overload. Be sure to use an exact replacement.

7. Check the continuity between points 4 and 5. If continuity is found, continue to make the following checks. If no continuity is found, repair the broken wire or loose connection.

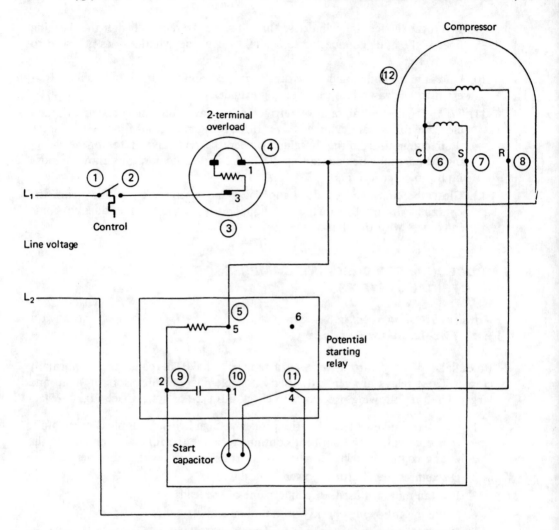

Figure 4-25. Potential starting relay system with two-terminal overload.

8. Check the continuity between points 4 and 6. If continuity is found, continue to make the following checks. If no continuity is found, repair the broken wire or loose connection.

9. Remove the wiring from the start terminal (point 7). Check the continuity between points 6 and 7. Compare this reading to that given by the compressor manufacturer for that particular compressor. If proper continuity is found, continue to make the following checks. If no or improper continuity is found, replace the compressor. The start winding is defective.

10. Remove the wiring from the run terminal (point 8). Check the continuity be-

tween points 6 and 8. Compare this reading to that given by the compressor manufacturer for that particular compressor. If proper continuity is found, continue to make the following checks. If no or improper continuity is found, replace the compressor. The run winding is defective.

11. Check the continuity between points 5 and 9. If continuity is found, continue to make the following checks. If no continuity is found, the start relay coil is open. Replace the relay. Be sure to use an exact replacement.

12. Check the continuity between points 9 and 10. If continuity is found, continue to make the following checks. If no continuity is found, the start relay contacts are defective. Replace the relay. Be sure to use an exact replacement.

13. Set the ohmmeter on R × 100,000. Check the continuity between points 10 and 11. If needle deflection is found, continue to make the following checks. If no needle deflection is found, the capacitor is open. Replace the capacitor. Be sure to use a proper replacement.

14. Set the ohmmeter on R × 1. Check the continuity between points 10 and 11. If no continuity is found, continue to make the following checks. If continuity is found, the capacitor is shorted. Replace the capacitor. Be sure to use a proper replacement.

15. Check the continuity between points 6 and 12. If no continuity is found, continue to make the following checks. If continuity is found, the compressor motor is shorted. Replace the compressor.

16. Check the continuity between point 9 and the wire removed from point 7 (start terminal). If continuity is found, replace the wiring on the terminal and continue to make the following checks. If no continuity is found, repair the wiring or loose connection.

17. Check the continuity between point 11 and the wire removed from point 8 (run terminal). If continuity is found, replace the wiring on the terminal and continue to make the following checks. If no continuity is found, repair the wiring or loose connection.

18. If the steps above do not indicate the trouble and the compressor will operate for a short period of time, disconnect the wire from point 10. Touch the wire to the same terminal and turn on the electricity. When the compressor starts, immediately remove the wire from the terminal. A slight spark may occur. If the compressor continues to operate, replace the start relay. The contacts are not opening. Be sure to use an exact replacement. If the compressor does not continue to run when the wire is removed, reconnect the wire and proceed to the next step.

19. Check the amperage draw through the wire to point 6 (common terminal) while the compressor is trying to run. Compare the amperage draw to the locked rotor (LR) amperage of the motor. If the LR amperage is very close to that listed by the manufacturer, replace the compressor. The compressor has mechanical problems.

Potential Starting Relay System
with Three-Terminal Overload

Potential (voltage) starting relays are nonpositional. The contacts are normally closed. These relays are generally used in single-phase compressors of ½ hp and larger. Use the following procedure to check this type of system (see Fig. 4–26):

1. Check to make sure that the proper voltage is being supplied to the unit. If the proper voltage is being supplied, continue to make the following checks. If the wrong voltage is being supplied, correct the problem with the power supply.

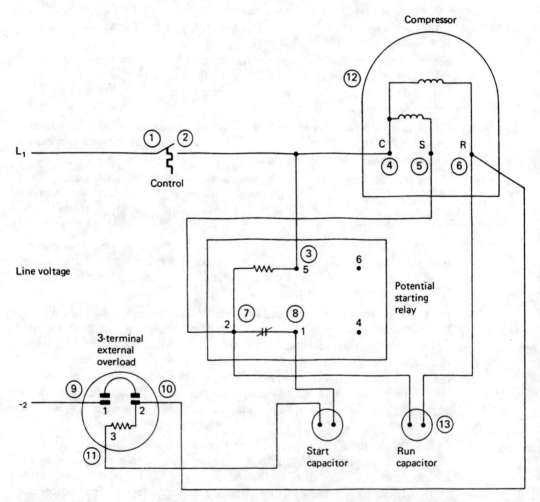

Figure 4–26. Potential relay systems with a three-terminal overload.

2. Disconnect the electrical power from the unit.

3. If a fan is used, disconnect one of the leads.

4. Make the following check with an ohmmeter.

5. Check the continuity between points 1 and 2. Refer to Figure 4-26. If continuity is found, continue to make the following checks. No continuity indicates the control contacts are open. Close the control contacts by setting the control to demand operation. If there is no continuity with the control on demand, replace the control.

6. Check the continuity between points 2 and 3. If continuity is found, continue to make the following checks. If no continuity is found, repair the broken wire or loose connection.

7. Check the continuity between points 2 and 4. If continuity is found, continue to make the following checks. If no continuity is found, repair the broken wire or loose connection.

8. Remove the wiring from point 5 (start terminal). Check the continuity between points 4 and 5. Compare this reading to that given by the compressor manufacturer for that particular compressor. If proper continuity is found, continue to make the following checks. If no or improper continuity is found, replace the compressor. The start winding is defective.

9. Remove the wiring from point 6 (run terminal). Check the continuity between points 4 and 6. Compare this reading to that given by the manufacturer for that particular compressor. If proper continuity is found, continue to make the following checks. If no or improper continuity is found replace the compressor. The run winding is defective.

10. Check the continuity between points 3 and 7. If continuity is found, continue to make the following checks. If no continuity is found, replace the relay. The coil is faulty. Be sure to use an exact replacement.

11. Check the continuity between points 7 and 8. If continuity is found, continue to make the following checks. If no continuity is found, replace the relay. The contacts are faulty. Be sure to use an exact replacement.

12. Set the ohmmeter on R × 100,000. Check the continuity between points 8 and 11. If needle deflection is found, continue to make the following checks. If no needle deflection is found, the start capacitor is open. Replace the capacitor. Be sure to use a proper replacement.

13. Set the ohmmeter on R × 1. Check the continuity between points 9 and 11. If no continuity is found, continue to make the following checks. If continuity is found, the start capacitor is shorted. Be sure to use a proper replacement.

14. Set the ohmmeter on R × 100,000. Check the continuity between points 7 and the wire removed from point 6 (run terminal). If needle deflection is found, continue to make the following checks. If no needle deflection is found, the run capacitor is open. Replace the capacitor. Be sure to use a proper replacement.

15. Set the ohmmeter on R × 1. Check the continuity between point 7 and the wire removed from point 6 (run terminal). If no continuity is found, continue to make the following checks. If continuity is found, the run capacitor is shorted. Replace the capacitor. Be sure to use a proper replacement.

16. Check the continuity between points 9 and 11. If continuity is found, continue to make the following checks. If no continuity is found, replace the overload. Be sure that the unit has been off for about 10 minutes to allow the overload to cool. Be sure to use an exact replacement.

17. Check the continuity between points 4 and 12. If no continuity is found, continue to make the following checks. If continuity is found, replace the compressor. The motor is grounded.

18. Check the continuity between point 7 and the wire removed from point 5 (start terminal). If continuity is found, continue to make the following checks. If no continuity is found, repair the wiring or loose connection. Reconnect the wiring to terminal (point 5).

19. Check the continuity between point 13 and the wire removed from point 6 (run terminal). If continuity is found, continue to make the following checks. If no continuity is found, repair the wiring or loose connection. Reconnect this wire to the terminal (point 6).

20. Check the continuity between point 10 and the wire removed from point 6 (run terminal). If continuity is found, continue to make the following checks. If no continuity is found, repair the wiring or loose connection. Reconnect this wire to the terminal (point 6).

21. If the steps above do not indicate the trouble and the compressor will operate for a short period of time, disconnect the wire from point 8. Touch the wire to the same terminal and turn on the electricity. When the compressor starts, remove the wire from the terminal. A slight spark may occur. If the compressor continues to operate, replace the start relay. The contacts are not operating. Be sure to replace it with an exact replacement. If the compressor does not continue to run when the wire is removed, reconnect the wire and proceed to the next step.

22. Check the amperage through the wire to point 4 (common terminal) while the compressor is trying to run. Compare the amperage draw to the locked rotor (LR) amperage of the motor. If the LR amperage is very close to that listed by the manufacturer, replace the compressor. The compressor has mechanical problems.

Current-Type Starting Relay System with Two-Terminal Overload

Current-type starting relays are positional-type relays. They must be mounted in the position indicated by arrows on the relay. The contacts are normally open, and they are closed by an electromagnetic coil. They are opened by gravity—thus the purpose of mounting in the proper position. These relays are generally used on fractional-horsepower motors up to about ½ hp. Use the following procedure to check this type of system (see Fig. 4–27):

1. Check to make certain that the proper voltage is being supplied to the unit. If proper voltage is being supplied, continue to make the following checks. If the wrong voltage is being supplied, correct the problem with the power supply.

2. Disconnect the electrical power from the unit.

3. If a fan motor is used, disconnect one of the leads.

4. Make the following check with an ohmmeter.

5. Check the continuity between points 1 and 2. Refer to Fig. 4–27. No continuity indicates that the control contacts are open. Close the contacts by setting the

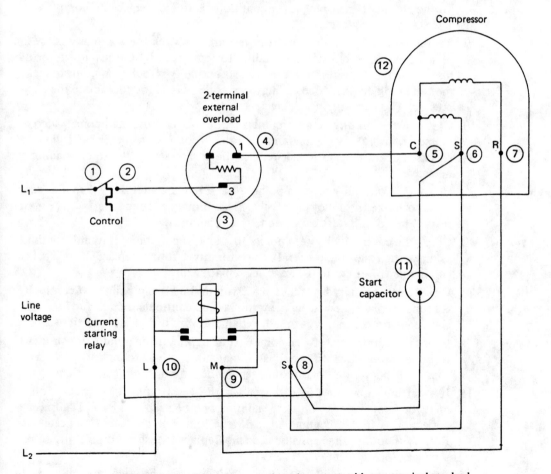

Figure 4–27.   Current-type starting relay system with two-terminal overload.

control to demand operation. If continuity is found, continue to make the following checks. If no continuity is found with the control on demand, replace the control.

6. Check the continuity between points 3 and 4. If continuity is found, continue to make the following checks. If no continuity is found, wait about 10 minutes and check the continuity again. The overload may be tripped. If no continuity is found now, replace the overload. Be sure to use an exact replacement.

7. Check the continuity between points 4 and 5. If continuity is found, continue to make the following checks. If no continuity is found, repair the wire or loose connections.

8. Remove the wire from point 6 (start terminal). Check for continuity between points 5 and 6. Compare this reading given by the manufacturer of that particular compressor. If proper continuity is found, continue to make the following checks. If no or improper continuity is found, replace the compressor. The start winding is defective.

9. Remove the wire from point 7 (run terminal). Check the continuity between points 5 and 7. Compare this reading to that given by the manufacturer of that particular compressor. If the proper continuity is found, continue to make the following checks. If no or improper continuity is found, replace the compressor. The run winding is defective.

10. Check the continuity between point 5 (common terminal) and point 12 (compressor housing). If no continuity is found, continue to make the following checks. If continuity is found, replace the compressor. The motor winding is defective.

11. Check the continuity between points 10 and 8. If no continuity is found, continue to make the following checks. If continuity is found, replace the start relay. The contacts are closed and should be open.

12. Check the continuity between points 10 and 9. If continuity is found, continue to make the following checks. If no continuity is found, replace the relay. The coil is open. Be sure to use an exact replacement.

13. Check the continuity between point 7 (run terminal) and point 9. If continuity is found, replace the wire on the terminal and continue to make the following checks. If no continuity is found, repair the wire or loose connections.

14. Check the continuity between point 6 (start terminal) and point 8. If continuity is found, replace the wire on the terminal and continue to make the following checks. If no continuity is found, repair the wire or loose connections.

15. If a starting capacitor is used, disconnect the wire from point 11. Set the ohmmeter on R × 100,000. Check the continuity between points 8 and 11. If needle deflection is found, continue to make the following checks. If no needle deflection is found, the capacitor is open. Replace the capacitor. Be sure to use a proper replacement.

16. Set the ohmmeter on R × 1. Check the continuity between points 8 and 11.

If no continuity is found, continue to make the following checks. If continuity is found, the capacitor is shorted. Replace the capacitor. Be sure to use a proper replacement.

17. If the steps above do not indicate the trouble and the compressor hums but does not start, disconnect the wire from point 8 and touch it to point 10. Turn on the electricity. If the compressor starts, immediately remove the wire from the terminal. A slight spark may occur. If the compressor continues to operate, replace the relay. Be sure to use an exact replacement. If the compressor does not start, reconnect the wire to point 8 and proceed to the next step.

18. Check the amperage in the wire to point 5 (common terminal) while the compressor is trying to run. Compare the amperage draw to the locked rotor (LR) amperage of the motor. If the LR is very close to that listed by the manufacturer, replace the compressor. The compressor has mechanical problems.

Permanent Split Capacitor System with Two-Terminal Overload

PSC motors are low-starting-torque motors that have no starting relay or starting capacitor. This system is used on builder's models and room-type air-conditioning units. Use the following procedure to check this type of system (see Fig. 4-28).

1. Check to make certain that the proper voltage is being supplied to the unit. If proper voltage is being supplied, continue to make the following checks. If the wrong voltage is being supplied, correct the problem with the power supply.

2. Disconnect the electrical power from the unit.

3. If a fan motor is used, disconnect one of the leads.

4. Make the following checks with an ohmmeter. Be sure to zero the ohmmeter.

5. Check the continuity between points 1 and 2. Refer to Fig. 4-28. If continuity is found, continue to make the following checks. No continuity indicates the control contacts are open. Close the contacts by setting the control to demand operation. If no continuity is found with the control on demand, replace the control.

6. Check the continuity between points 2 and 3. If continuity is found, continue to make the following checks. If no continuity is found, repair the wiring or loose connections.

7. Remove the wire from point 4 (start terminal). Check the continuity between points 3 and 4. Compare this reading to that given by the manufacturer for that particular compressor. If proper continuity is found, continue to make the following checks. If no or improper continuity is found, replace the compressor. The start winding is defective.

8. Remove the wiring from point 5 (run terminal). Check the continuity between

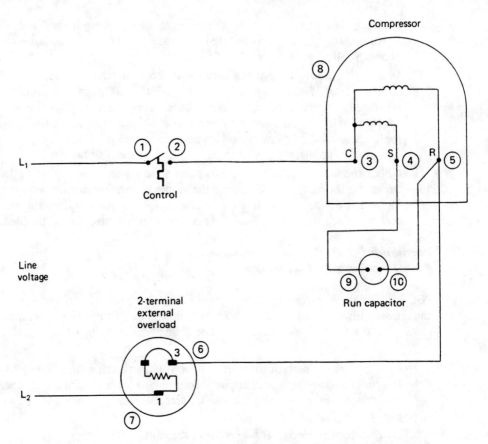

Figure 4-28. Permanent split capacitor (PSC) system with two-terminal overload.

points 3 and 5. Compare this reading to that given by the manufacturer for that particular compressor. If proper continuity is found, continue to make the following checks. If no or improper continuity is found, replace the compressor. The run winding is defective.

9. Check the continuity between point 3 (common terminal) and point 8 (compressor housing). If no continuity is found, continue to make the following checks. If continuity is found, replace the compressor. The motor winding is grounded.

10. Set the ohmmeter on R × 100,000. Check the continuity between points 9 and 10. If needle deflection is found, continue to make the following checks. If no needle deflection is found, the capacitor is open. Replace the capacitor. Be sure to use a proper replacement.

11. Set the ohmmeter on R × 1. Check the continuity between points 9 and 10. If no continuity is found, continue to make the following checks. If continuity

is found, the capacitor is shorted. Replace the capacitor. Be sure to use a proper replacement.

12. Check the continuity between points 5 and 6. If continuity is found, continue to make the following checks. If no continuity is found, repair the wire or loose connection.

13. Check the continuity between points 6 and 7. If continuity is found, continue to make the following checks. If no continuity is found, wait about 10 minutes and check the continuity again. The overload may be tripped. If no continuity is found, replace the overload. Be sure to use an exact replacement.

14. Reconnect all wiring. Connect a 12-in. piece of electrical wire to each terminal on a 130 μF × 370 V ac start capacitor. Touch the loose ends of one wire to point 4 (start terminal) and the other wire to point 5 (run terminal). Turn on the electricity. If the compressor starts, immediately remove the two wires from points 4 and 5. If the compressor continues to run, install a hard-start kit (see Fig. 4-29). If the compressor does not start or stops running when the wires are removed, proceed to the next step.

15. Check the amperage draw in the wire to point 3 (common terminal) while the compressor is trying to start. Compare the amperage to the locked rotor (LR) amperage of the motor. If the LR amperage is very close to that listed by the manufacturer, replace the compressor. The compressor has mechanical problems.

Permanent Split Capacitor System with Internal Thermostat Overload

PSC motors are low-starting-torque motors that have no starting relay or starting capacitor. However, they do have a run capacitor. These systems have no external overload because a more sensitive internal overload is located inside the compressor housing. This system is used on builder's models and room air-conditioning units. Use the following procedure to check this type of system (see Fig. 4-30):

1. Check to make certain that the proper voltage is being supplied to the unit. If proper voltage is being supplied, continue to make the following checks. If the wrong voltage is being supplied, correct the problem with the power supply.

2. Disconnect the electrical power from the unit.

3. If a fan motor is used, disconnect one of the leads.

4. Make the following checks with an ohmmeter. Be sure to zero the ohmmeter.

5. Check the continuity between points 1 and 2. Refer to Fig. 4-30. If continuity is found, continue to make the following checks. No continuity indicates the control contacts are open. Close the contacts by setting the control to demand operation. If no continuity is found with the control on demand, replace the control.

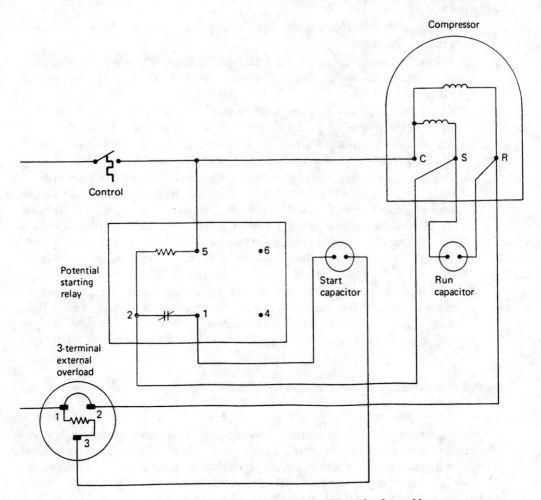

Figure 4–29. Connections for installing a hard-start kit.

6. Check the continuity between points 2 and 3. If continuity is found, continue to make the following checks. If no continuity is found, repair the wiring or loose connections.

7. Remove the wire from point 4 (start terminal). Check the continuity between points 3 and 4. Compare this reading to that given by the compressor manufacturer for that particular compressor. If continuity is found, continue to make the following checks. If no or improper continuity is found, remove the wire from point 5 (run terminal). Check the continuity between points 4 and 5. If no continuity is found, it can be assumed that the start winding is open. Replace the compressor. Check the continuity between points 3 and 5, if no continuity is found, the internal overload may be tripped. Cool the compressor

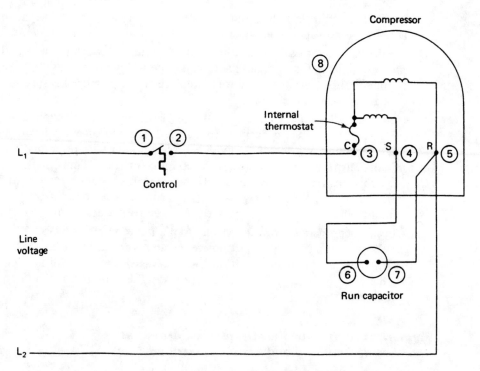

Figure 4-30. Permanent split capacitor (PSC) system with internal thermostat.

shell below 125°F (51.78°C) or until the hand may be held on the compressor without much discomfort. Again check the continuity between points 3 and 4. If no continuity is found, the internal overload is defective. Replace the compressor.

8. Remove the wire from point 5 (run terminal). Check the continuity between points 3 and 5. Compare this reading to that given by the manufacturer for that particular compressor. If proper continuity is found, continue to make the following checks. If no or improper continuity is found, check the continuity between points 4 and 5. If continuity is found, it can be assumed that the internal overload is open. Cool the compressor shell below 125°F (51.78°C) or until the hand can be held on the compressor without much discomfort. Again check the continuity between points 3 and 5. If no continuity is found, the internal overload is defective. Replace the compressor.

9. Check the continuity between points 3 and 8. If no continuity is found, continue to make the following checks. If continuity is found, replace the compressor. The motor winding is grounded.

10. Set the ohmmeter on R × 100,000. Check the continuity between points 6 and 7. If needle deflection is found, continue to make the following checks. If no

needle deflection is found, the capacitor is open. Replace the capacitor. Be sure to use a proper replacement.

11. Set the ohmmeter on R × 1. Check the continuity between points 6 and 7. If no continuity is found, continue to make the following checks. If continuity is found, the capacitor is shorted. Replace the capacitor. Be sure to use a proper replacement.

12. Reconnect all wiring. Connect a 12-in. piece of wire to each terminal on a 130 μF × 370 V ac start capacitor. Touch the loose end of one wire to point 4 (start terminal) and touch the loose end of the other wire to point 5 (run terminal). Turn on the electricity. If the compressor starts, immediately remove the two wires from points 4 and 5. If the compressor continues to run, install a hard-start kit. Refer to Fig. 4–29. If the compressor does not start or stops running when the wires are removed, proceed to the next step.

13. Check the amperage draw in the wire to point 3 (common terminal) while the compressor is trying to run. Compare the amperage draw to the locked rotor (LR) amperage of the motor. If the LR amperage is very close to that indicated by the manufacturer, replace the compressor. The compressor has mechanical problems.

Permanent Split Capacitor System with Internal Thermostat Overload and External Overload

PSC motors are low-starting-torque motors that have no starting relay or starting capacitor. However, they do use a run capacitor. This particular compressor has both an internal thermostat and an external overload for added protection. The internal thermostat is temperature-sensitive and interrupts the line voltage to the compressor motor common terminal. The external overload is both temperature-sensitive and current-sensitive. Its contacts interrupt the control circuit. Use the following procedure to check this type of system (see Fig. 4–31):

1. Check to make certain that the proper voltage is being supplied to the unit. If proper voltage is being supplied, continue to make the following checks. If the wrong voltage is being supplied, correct the problem with the power supply.

2. Disconnect the electrical power from the unit.

3. If a fan motor is being used, disconnect one lead.

4. Make the following checks with an ohmmeter. Be sure to zero the ohmmeter.

5. Check the continuity between points 1 and 2. Refer to Fig. 4–31. If continuity is found, continue to make the following checks. No continuity indicates that the overload is open. Wait about 10 minutes. Check the continuity again. If no continuity now, replace the external overload. Be sure to use an exact replacement.

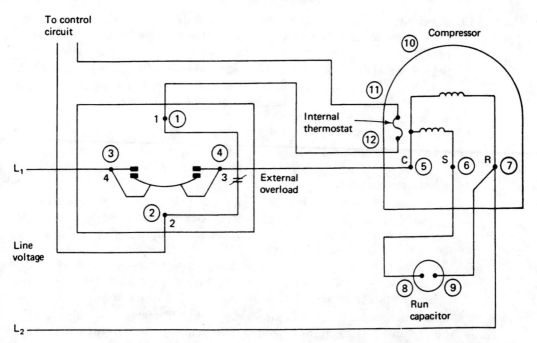

Figure 4–31. Permanent split capacitor (PSC) system with internal thermostat and external overload.

6. Check the continuity between points 3 and 4. If continuity is found, continue to make the following checks. No continuity indicates the external overload is defective. Replace the overload. Be sure to use an exact replacement.

7. Remove the wire from point 6 (start terminal). Check the continuity between points 5 and 6. Compare this reading to that given by the manufacturer for that particular compressor. If proper continuity is found, continue to make the following checks. If no or improper continuity is found, the start winding is defective. Replace the compressor.

8. Remove the wire from point 7 (run terminal). Check the continuity between points 5 and 7. Compare this reading to that given by the manufacturer for that particular compressor. If proper continuity is found, continue to make the following checks. If no or improper continuity is found, the run winding is defective. Replace the compressor.

9. Remove the wire from point 11. Check the continuity between points 11 and 12. If continuity is found, continue to make the following checks. If no or improper continuity is found, the internal thermostat is open. Cool the compressor below 125°F (51.78°C) or until the hand may be held on the compressor without much discomfort. Again check the continuity between points 11 and 12. If no continuity is found, replace the compressor. The internal thermostat is defective.

10. Check the continuity between points 5 and 10. If no continuity is found, continue to make the following checks. If continuity is found, replace the compressor. The motor winding is grounded.

11. Set the ohmmeter on R × 100,000. Check the continuity between points 8 and 9. If needle deflection is found, continue to make the following checks. If no needle deflection is found, the capacitor is open. Replace the capacitor. Be sure to use a proper replacement.

12. Set the ohmmeter on R × 1. Check the continuity between points 8 and 9. If no continuity is found, continue to make the following checks. If continuity is found, the capacitor is shorted. Replace the capacitor. Be sure to use a proper replacement.

13. Reconnect all wiring. Connect a 12-in. piece of electrical wire to each terminal on a 130 μF × 370 V ac start capacitor. Touch the loose end of one wire to point 6 (start terminal) and the loose end of the other wire to point 7 (run terminal). Turn on the electricity. If the compressor starts, immediately remove the two wires from points 6 and 7. If the compressor continues to run, install a hard-start kit (see Fig. 4–29). If the compressor does not start or stops running when the wires are removed, proceed to the next step.

14. Check the amperage draw in the wire to point 5 (common terminal) while the compressor is trying to run. Compare the amperage draw to the locked rotor (LR) amperage of the motor. If the LR amperage is very close to that indicated by the manufacturer, replace the compressor. The compressor has mechanical problems.

4-16 PROCEDURE FOR CHECKING THERMOSTATIC EXPANSION VALVES*

In checking complaints, if the expansion valve is suspected as the source of trouble, an orderly procedure for locating the exact difficulty and remedying it in the shortest time possible is desirable.

1. Check the suction pressure and discharge pressure. They act as the "pulse" of the refrigerating system and are the best guide to locating trouble.

2. Check the superheat. For this purpose, an accurate pocket thermometer may be taped to the suction line at the remote bulb location.

The following headings group the suggested possibilities of trouble indicated by the gauge and superheat readings.

*This section reprinted by permission of Alco Valve Co., St. Louis, Missouri.

Low Suction Pressure—High Superheat

Expansion valve limiting flow

1. Inlet pressure too low from excessive vertical lift, too small liquid line, or excessively low condensing temperature. Resulting pressure difference across valve too small.
2. Gas in liquid line—due to pressure drop in the line or insufficient refrigerant charge. If there is no sight glass in the liquid line, a characteristic whistling noise may be observed at the expansion valve.
3. Valve restricted by pressure drop through coil, requiring change to external equalizer (see Fig. 4–32).
4. External equalizer line plugged or external equalizer connection capped without providing a new valve cage or body with internal equalizer.
5. Moisture, wax, oil, or dirt plugging valve orifice. Ice formation or wax at valve seat may be indicated by sudden rise in suction pressure after shutdown and system has warmed up.
6. Valve orifice too small.
7. Superheat adjustment too high.
8. Power assembly failure or partial loss of charge.
9. Remote bulb of gas-charged thermo expansion valve has lost control due to the remote bulb tubing or power head being colder than the remote bulb.

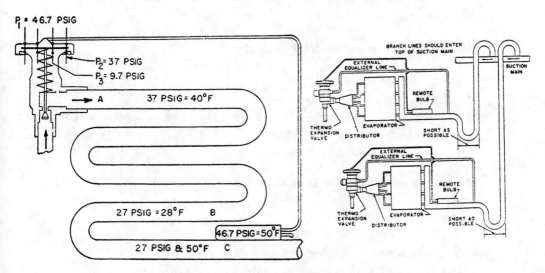

Figure 4-32.   External equalizer connecting methods. (Courtesy of Alco Controls Division, Emerson Electric Co.)

10. Filter screen clogged.

11. Wrong type of oil.

Restriction in system other than thermo valve. (Usually, but not necessarily, indicated by frost or lower-than-normal temperature at point of restriction)

1. Strainers clogged or too small.
2. Solenoid valve failure or valve undersized.
3. King valve at liquid receiver outlet too small or not fully opened.
4. Plugged lines.
5. Hand valve stem failure or valve too small or not fully opened.
6. Liquid line too small.
7. Suction line too small.
8. Wrong type of oil in system, blocking liquid flow.
9. Discharge or suction service valve on compressor restricted or not fully opened.

Low Suction Pressure—Low Superheat

1. Poor distribution in evaporator, causing liquid to short circuit through favored passes, throttling valve before all passes receive sufficient refrigerant.
2. Compressor oversized or running too fast due to wrong pulley.
3. Uneven or inadequate coil loading, poor air distribution or brine flow.
4. Evaporator too small—indicated sometimes by excessive ice formation.
5. Evaporator oil logged.

High Suction Pressure—High Superheat

1. Compressor undersized.
2. Evaporator too large.
3. Unbalanced system having an oversized evaporator, an undersized compressor and a high load on the evaporator.
4. Compressor discharge valves leaking.

High Suction Pressure—Low Superheat

1. Compressor undersized.
2. Valve superheat setting too low.
3. Gas in liquid line with oversized thermo valve.
4. Compressor discharge valves leaking.
5. Pin and seat of expansion valve wire drawn, eroded, or held open by foreign material resulting in liquid flood-back.

6. Ruptured diaphragm or bellows in a constant pressure (automatic) expansion valve, resulting in liquid flood-back.

7. External equalizer line plugged or external equalizer connection capped without providing a new valve cage or body with internal equalizer.

8. Moisture freezing valve in open position.

Fluctuating Suction Pressure

Poor control

1. Improper superheat adjustment.
2. Trapped suction line (see Fig. 4-33).
3. Improper remote bulb location or application (see Fig. 4-34).
4. "Flood-back" of liquid refrigerant caused by poorly designed liquid distribution device or uneven coil circuit loading. Also improperly hung evaporator. Evaporator not plumb (see Fig. 4-35).
5. External equalizer tapped at common point on application with more than one valve on same evaporator.
6. Faulty condensing water regulator, causing change in pressure drop across valve.
7. Evaporative condenser cycling, causing radical change in pressure difference across expansion valve.
8. Cycling of blowers or brine pumps.

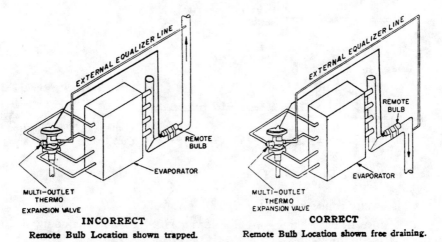

INCORRECT
Remote Bulb Location shown trapped.

CORRECT
Remote Bulb Location shown free draining.

Figure 4-33. Remote bulb installed on a trapped suction line. (Courtesy of Alco Controls Division, Emerson Electric Co.)

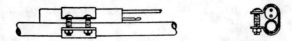

EXTERNAL BULB ON SMALL SUCTION LINE

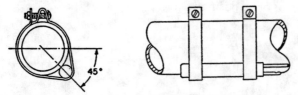

EXTERNAL BULB ON LARGE SUCTION LINE

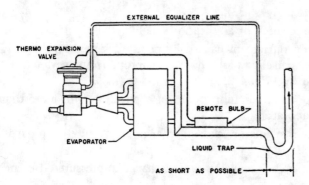

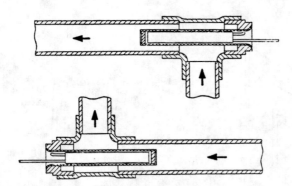

Figure 4–34. Recommended remote bulb location and schematic piping for rising suction line. (Courtesy of Alco Controls Division, Emerson Electric Co.)

Fluctuating Discharge Pressure

1. Faulty condensing water regulating valve.
2. Insufficient charge—usually accompanied by corresponding fluctuation in suction pressure.
3. Cycling of evaporative condenser.
4. Inadequate and fluctuating supply of cooling water to condenser.

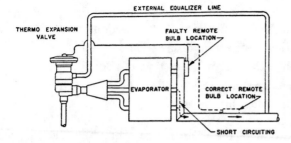

Figure 4-35. Correct remote bulb location on a "short-circuiting" evaporator to prevent "flood-back." (Courtesy of Alco Controls Division, Emerson Electric Co.)

High Discharge Pressure

1. Insufficient cooling water. (Inadequate supply or faulty water valve.)
2. Condenser or liquid receiver too small.
3. Cooling water above design temperature.
4. Air or noncondensable gases in condenser.
5. Overcharge of refrigerant.
6. Condenser dirty.
7. Air-type condenser improperly located to dispel hot discharge air.

Expansion Valve "Freeze-Ups"

Expansion valve trouble may occur from the formation of ice crystals or the separation of wax out of the oil at the pin and seat, causing the refrigerant flow to be restricted or stopped entirely. Waxing generally occurs only at extreme low temperatures; however, ice formation due to moisture in the system can occur whenever the evaporator is operating below freezing. Moisture is generally indicated by the starving of the evaporator. However, it may cause a valve to freeze open and cause "flood-back." When the compressor stops and the expansion valve and the system are allowed to warm up, the system will operate satisfactorily until the ice crystals form again. Another indication of moisture in the expansion valve can be noted if tapping the valve body will cause it to feed again for a short time. Under no circumstances should a torch be used on the valve body to melt ice accumulation at the pin and seat. If the ice must be melted to put the system in operation temporarily, then apply hot rags. This is equally effective and will not damage the valve.

To remedy this, install a dryer of suitable size and operate the system above freezing for a few days to allow the moisture to be trapped in the dryer rather than to freeze at some location in the evaporator. If moisture persists, then the refrigerant charge and the oil should be dumped, and the system dried by means of a vacuum and heat or by blowing dry nitrogen gas through the system.

Moisture can be admitted to the system:

1. At the time of original installation from moist air in the piping.

2. At the time of charging by not blowing the air and moisture out of the charging hose before tightening the fitting at the connection.
3. From improperly refilled refrigerant drums.
4. By allowing air to enter the system at the time of adding oil.
5. From leaking or defective shaft seal.
6. By opening the system when in a vacuum.

In a methyl chloride system, regardless of the operating range, the presence of moisture creates an acid condition that attacks the system, creates sludge, and causes copper to plate upon the steel parts of the compressor as well as the pin and seat of the expansion valve.

4-17 PROCEDURE FOR TORCH BRAZING*

In this section we describe procedures that are important for making sound brazements of tubing in the air-conditioning and refrigeration industry, using the phoscopper and the silver brazing filler metals. The importance of brazing cannot be overlooked and/or overemphasized since it is a major part of the air-conditioning and refrigeration industry.

Brazing is the process used in joining the major components of a refrigeration system into a closed circuit. Since the closed circuit contains a refrigerant, every brazed joint must be leak-free; if not, the refrigerant will escape, creating a severe inconvenience to the customer as well as a costly repair. The purpose of this section is to provide the basic information about the correct method for torch brazing.

Definitions

Brazing is the application of heat above a temperature of 800°F and below the melting points of the base metals to produce coalescence or bonding by surface adhesion forces between the molten filler metals and the surfaces of the base metals. The filler metal is distributed through the joint by using capillary action. Brazing is not to be confused with soldering, even though the procedures are very similar. Soldering is a term used for metal-joining processes at temperatures below 800°F.

For bonding and distribution by capillary attraction to occur, the filler metal must be able to "wet" the base metals. Wetting is the phenomenon in which the forces of attraction between the molecules of the molten filler metal and the molecules of the base metals are greater than the inward forces of attraction existing between the molecules of the filler metal. The degree of wetting is a function of the compositions of the base metal and the filler metal and of the temperature. Good wetting can only occur on perfectly clean and oxide-free surfaces.

*This section by Engelhard Industries, Warwick, Rhode Island. Used with permission.

Filler Metals

The quality and strength of a brazement are more a function of the physical parameters of the joint and the brazing procedures used than which filler metal is applied to the joint. These parameters determine the selection of the best suited and the most easily handled filler metal for a particular joint.

The phos-copper brazing alloys were specifically developed for joining copper and copper alloys. They are used for brazing copper, brass, bronze, or combinations of these. When brazing with either brass or bronze, a flux must be used to prevent the formation of an "oxide coat" over these base metals. This coat would prevent wetting and the flow of the filler metal. However, when brazing copper to copper joints, these alloys are self-fluxing. Because of the phosphorus embrittlement of the joint, phos-copper filler metals should not be used on ferrous metals or base metals containing more than 10% nickel. These alloys are also not recommended for use on aluminum bronze.

Unlike the phos-copper alloys, the silver brazing alloys contain no phosphorus. These filler metals are used to braze all ferrous, copper, and copper-based metals, with the exception of aluminum and magnesium. They require a flux with all applications. Extreme care should be taken when using the low-temperature, cadmium-bearing silver alloy because of poisonous cadmium fumes being emitted.

Most refrigeration repair work can be accomplished with two filler metals. Braze-it 15 (15% silver content) is a commonly used phos-copper filler metal, and Braze-it 45 w/cd (45% silver, cadmium-bearing) is a commonly used silver brazing filler metal.

Torch Brazing Procedures

Copper-to-copper tube joints using phos-copper brazing alloys

1. Correct oxygen and fuel gas mixture (see Fig. 4-36).

a. *Excessive gas mixture.* A reducing type of flame denotes excessive fuel gas; a greater amount of fuel gas than oxygen. A slightly reducing flame heats and cleans the metal surface for quicker and better brazing (see Fig. 4-37).

b. *Balanced gas mixture.* The gas mixture contains an equal amount of oxygen and fuel gas. It produces a flame that heats the metal, but has no other effect (see Fig. 4-38).

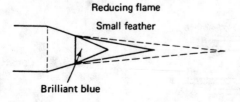

Reducing flame

Small feather

Brilliant blue

Figure 4-36.   Best type of flame for brazing. (Courtesy of Engelhard Corporation.)

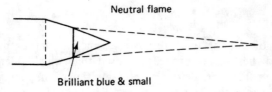

Neutral flame

Brilliant blue & small

Figure 4-37. Feather or reducing flame disappeared. (Courtesy of Engelhard Corporation.)

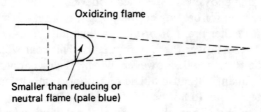

Oxidizing flame

Smaller than reducing or neutral flame (pale blue)

Figure 4-38. Worst type of flame for brazing. (Courtesy of Engelhard Corporation.)

c. *Excessive oxygen mixture.* The gas mixture contains an excessive amount of oxygen. It produces a flame that oxidizes the metal surface. A black oxide scale will form on the metal surface.

2. Cleanliness. General cleanliness is of prime importance to reliable brazing. All metal surfaces to be brazed must be cleaned of all dirt and foreign matter. When repair work is to be done, the metal surfaces to be joined must be either wire brushed and/or cleaned with sandpaper. A concerted effort must be made to keep oil, paint, dirt, grease, and aluminum off the surface of metals to be joined for these contaminants will:

a. Keep brazing filler metal from flowing into the joint.
b. Prevent brazing filler metal from wetting or bonding to the metal surfaces.

3. Correct insertion and clearance of parts. The minimum insertion distance is equal to the diameter of the inner tube. There should be 0.001 to 0.005 in. clearance between the walls of the inner and outer tubes (see Fig. 4-39).

4. Correct amount and application of heat. Heat evenly over the entire joint circumference, and joint length. Use enough heat to get entire joint hot enough to melt the brazing filler metal without heating the filler metal directly with the flame (see Fig. 4-40). Heat both tubes at the joint, and distribute heat evenly. *Do not overheat* the joint to the point that the metal to be joined begins to melt (see Fig. 4-41).

0.001"-0.005" clearance

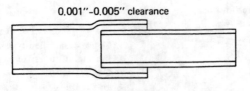

Figure 4-39. Correct clearance on joint parts. (Courtesy of Engelhard Corporation.)

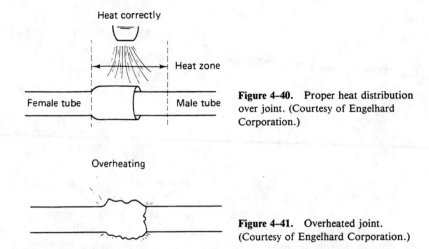

Figure 4–40. Proper heat distribution over joint. (Courtesy of Engelhard Corporation.)

Figure 4–41. Overheated joint. (Courtesy of Engelhard Corporation.)

Always try to use the correct size torch tip, and a slightly reducing flame. Overheating enhances the base metal-filler metal interactions (e.g., formation of chemical compounds). In the long run, these interactions are detrimental to the life of the joint. Some examples of heating methods are shown in Figs. 4–42 through 4–44.

5. Correct application of brazing rod (see Fig. 4–45). Use the brazing rod as a temperature indicator. If the brazing rod melts when placed on the hot joint, the tubing is hot enough to begin brazing. For best results, preheat the rod slightly with

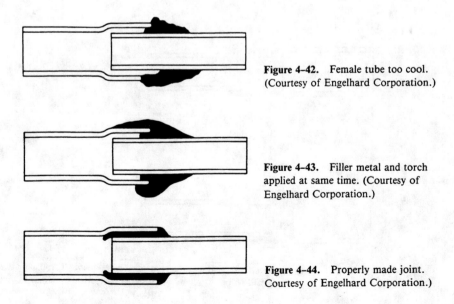

Figure 4–42. Female tube too cool. (Courtesy of Engelhard Corporation.)

Figure 4–43. Filler metal and torch applied at same time. (Courtesy of Engelhard Corporation.)

Figure 4–44. Properly made joint. Courtesy of Engelhard Corporation.)

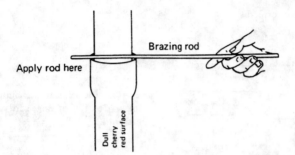

Figure 4–45. Correct application of brazing rod. (Courtesy of Engelhard Corporation.)

the outer envelope of the flame, but the hot copper tubing (not the direct flame) should melt the brazing rod.

6. Capillary attraction (see Fig. 4–46). This is the phenomenon by which the brazing filler metal is drawn into the joint. It is caused by the attraction between the molecules of the brazing filler metal and the molecules of the metal surfaces to be joined. It can only work if: (1) the surface of the metal is clean, (2) the clearance

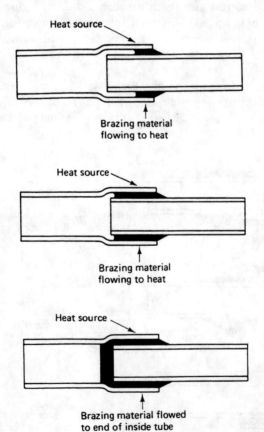

Figure 4–46. Capillary attraction. (Courtesy of Engelhard Corporation.)

between the metal surfaces is correct, (3) the metal at the joint area is hot enough to melt the brazing filler metal, and (4) the brazing filler metal will flow toward the heat source as illustrated.

Copper-to-brass tube joints using the phos-copper brazing alloys

1. The above-mentioned procedures for copper to copper joints are strictly followed.
2. Before application of heat to the joint, a small amount of flux is applied to allow wetting of the brass by the filler metal.
3. Upon completion of the brazement, the flux residue is thoroughly cleaned off. Cleaning can be accomplished with hot water and mechanical brushing. Most fluxes are corrosive and must be completely removed from the joint.

Steel-to-steel, copper, brass, or bronze using silver brazing alloys

1. The above-mentioned procedures for joining copper to copper are strictly followed.
2. Before application of heat to the joint, flux is applied to allow wetting and flow to take place between the filler metal and the base metals.
3. Before application of the filler metal to the joint, a small amount of flux is applied to the rod. Rod is heated up and then dipped into the flux. This coats the filler metal with a thin layer of flux that provides for quick flow by preventing the formation of an oxide coat around the filler metal (zinc oxide).
4. Flux must be completely washed off upon completion of the joint.

Fluxes

A flux is analogous to a sponge absorbing water, only a flux is absorbing oxides. It can only absorb so much before it becomes useless as a flux. When a flux becomes saturated with oxides, its viscosity increases. The ability of the flux to be displaced by the filler metal is therefore impeded. This leads to flux entrapment in the joint, which eventually causes corrosion and leaks.

One should use the least amount of flux that will get the job done. Then thoroughly clean the flux residue off after completing the brazement. Apply the flux along the surface of the joint and not into the joint itself. Allow the flux to flow into the joint ahead of the filler metal.

Simplified Rules for Good Brazing

1. Use a slightly reducing flame to get the maximum in heating and cleansing action.
2. Make sure metal surfaces are clean.

3. Examine parts for correct insertion and clearances.

4. Flux parts: (1) use minimum amount and (2) apply to outside of joint. No flux required for joining copper to copper with phos-copper filler metals.

5. Apply heat evenly around the joint to the proper brazing temperature.

6. Apply filler metal to joint. Make sure that it is completely distributed throughout the joint by using the torch. Molten filler metals will follow the heat.

7. If flux is used, thoroughly clean off the residue.

8. A major part of good brazing is to complete the brazement as quickly as possible. Keep the heating cycle short. Avoid overheating.

9. Have adequate ventilation. Brazing may result in hazardous fumes: cadmium fumes from cadmium-bearing filler metals and fluoride fumes from fluxes.

The following diagrams are taken from Lennox Industries and Magic Chef Wiring Diagrams.

LENNOX INDUSTRIES WIRING DIAGRAMS

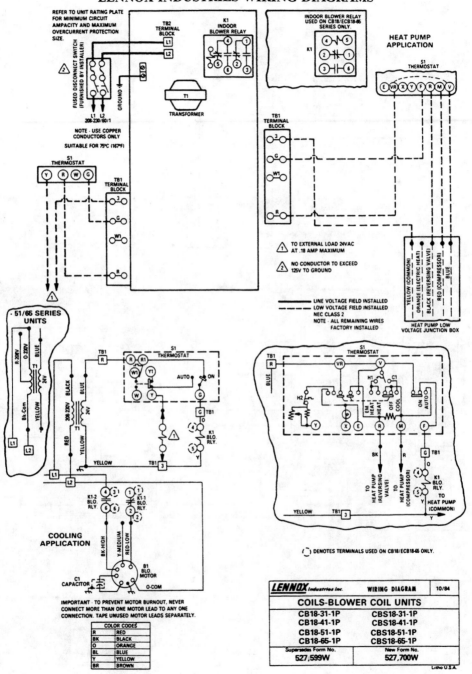

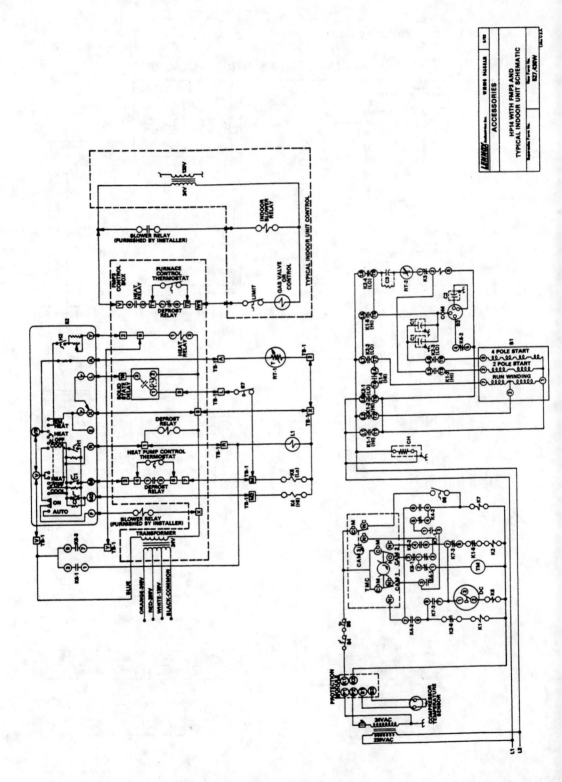

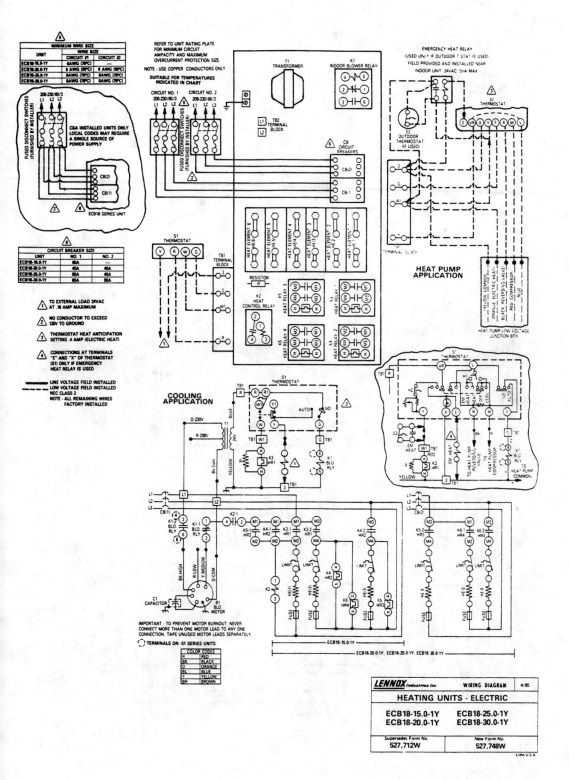

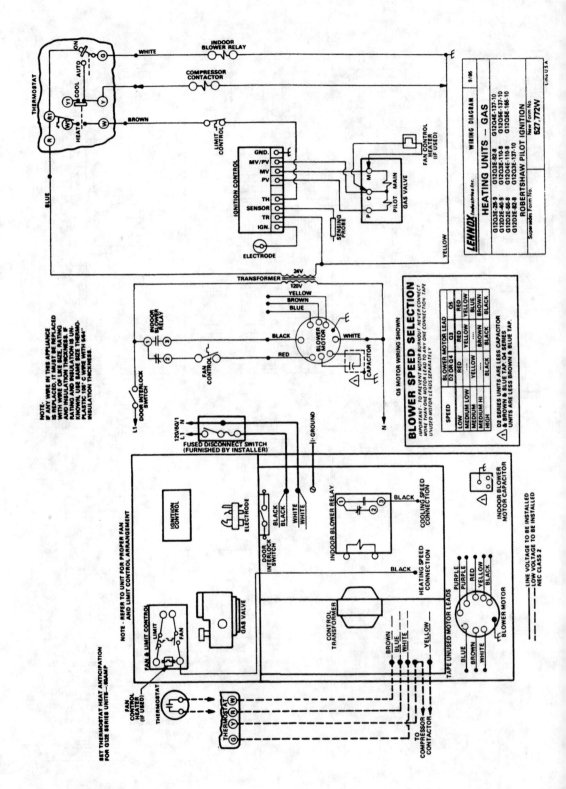

214

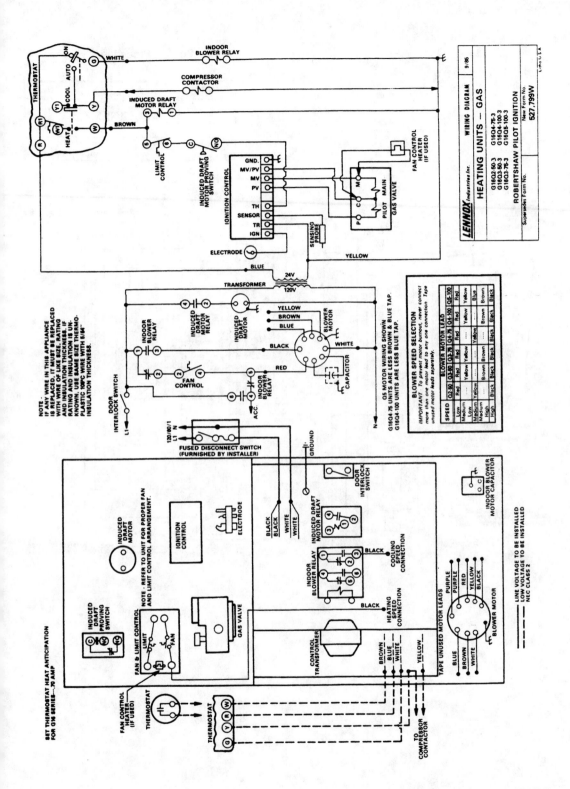

215

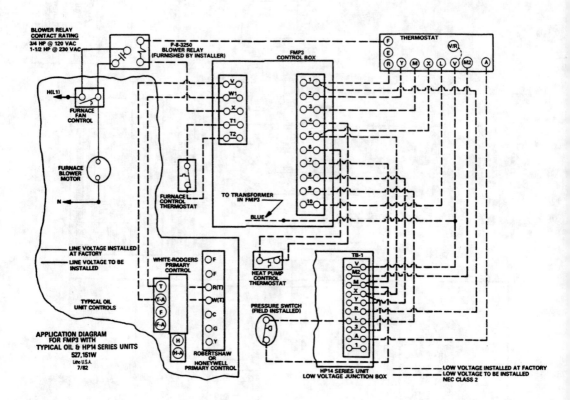

BLOWER RELAY
CONTACT RATING
3/4 HP @ 120 VAC
1-1/2 HP @ 230 VAC

P-8-3250
BLOWER RELAY
(FURNISHED BY INSTALLER)

FMP3
CONTROL BOX

THERMOSTAT

H(L1)

FURNACE FAN
CONTROL

FURNACE
BLOWER
MOTOR

N

FURNACE L
CONTROL
THERMOSTAT

TO TRANSFORMER
IN FMP3

BLUE

LINE VOLTAGE INSTALLED
AT FACTORY

LINE VOLTAGE TO BE
INSTALLED

TYPICAL OIL
UNIT CONTROLS

WHITE-RODGERS
PRIMARY
CONTROL

HEAT PUMP
CONTROL
THERMOSTAT

PRESSURE SWITCH
(FIELD INSTALLED)

TB-1

APPLICATION DIAGRAM
FOR FMP3 WITH
TYPICAL OIL & HP14 SERIES UNITS
527,151W
Litho U.S.A.
7/82

ROBERTSHAW
OR
HONEYWELL
PRIMARY CONTROL

HP14 SERIES UNIT
LOW VOLTAGE JUNCTION BOX

LOW VOLTAGE INSTALLED AT FACTORY
LOW VOLTAGE TO BE INSTALLED
NEC CLASS 2

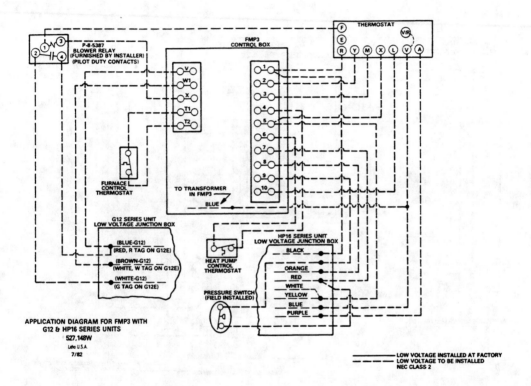

APPLICATION DIAGRAM FOR FMP3 WITH
G12 & HP16 SERIES UNITS
527,148W
Litho U.S.A.
7/82

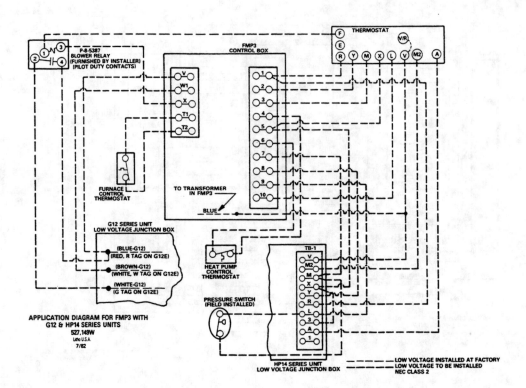

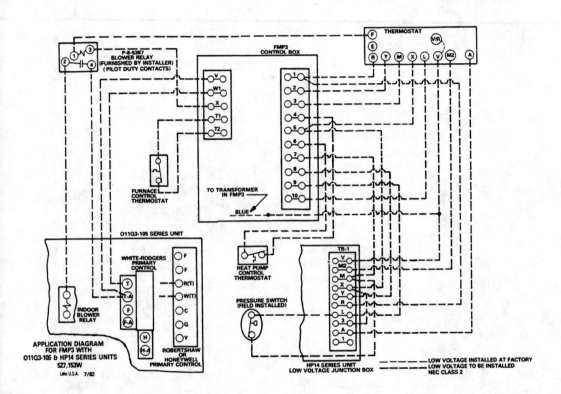

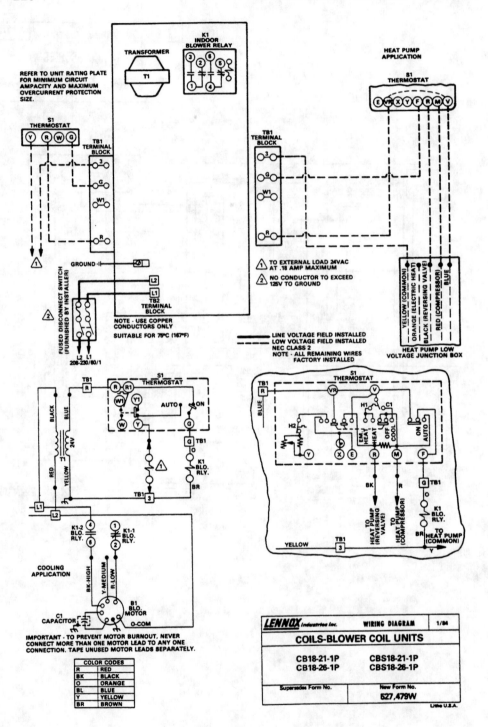

REFER TO UNIT RATING PLATE
FOR MINIMUM CIRCUIT
AMPACITY AND MAXIMUM
OVERCURRENT PROTECTION
SIZE.

TRANSFORMER

K1
INDOOR
BLOWER RELAY

T1

HEAT PUMP
APPLICATION

S1
THERMOSTAT

E VR X Y F R M V

S1
THERMOSTAT

Y R W G

TB1
TERMINAL
BLOCK

3
G
W1
R

TB1
TERMINAL
BLOCK

3
G
W1
R

GROUND

L2
L1
TB2
TERMINAL
BLOCK

1 TO EXTERNAL LOAD 24VAC
AT .18 AMP MAXIMUM

2 NO CONDUCTOR TO EXCEED
125V TO GROUND

FUSED DISCONNECT SWITCH
(FURNISHED BY INSTALLER)

NOTE - USE COPPER
CONDUCTORS ONLY
SUITABLE FOR 75°C (167°F)

L2 L1
208-230/60/1

LINE VOLTAGE FIELD INSTALLED
LOW VOLTAGE FIELD INSTALLED
NEC CLASS 2
NOTE - ALL REMAINING WIRES
FACTORY INSTALLED

YELLOW (COMMON)
ORANGE (ELECTRIC HEAT)
BLACK (REVERSING VALVE)
RED (COMPRESSOR)
BLUE

HEAT PUMP LOW
VOLTAGE JUNCTION BOX

TB1
R

S1
THERMOSTAT

R R1
W1 Y1
AUTO   ON
W   Y   G   TB1

BLACK   BLUE

RED   YELLOW

24V
T1

L1
L2

K1-2
BLO.
RLY.
6

1
2
K1-1
BLO.
RLY.

COOLING
APPLICATION

BK-HIGH   Y-MEDIUM   R-LOW

C1
CAPACITOR

B1
BLO.
MOTOR
O-COM

IMPORTANT - TO PREVENT MOTOR BURNOUT, NEVER
CONNECT MORE THAN ONE MOTOR LEAD TO ANY ONE
CONNECTION. TAPE UNUSED MOTOR LEADS SEPARATELY.

1
K1
BLO.
RLY.

TB1
3
BR

TB1
R

S1
THERMOSTAT

VR   V
H1   C1

BLUE

H2
EM.
HEAT
Y   X   E   R   M   F
HEAT OFF COOL ON AUTO

BK   R
HEAT PUMP
(REVERSING
VALVE)

TO
HEAT PUMP
(COMPRESSOR)

G TB1
K1
BLO.
RLY.
BR   TO
HEAT PUMP
(COMMON)
Y

YELLOW   TB1
3

COLOR CODES	
R	RED
BK	BLACK
O	ORANGE
BL	BLUE
Y	YELLOW
BR	BROWN

LENNOX Industries Inc.	WIRING DIAGRAM	1/84
COILS-BLOWER COIL UNITS		
CB18-21-1P CB18-26-1P	CBS18-21-1P CBS18-26-1P	
Supersedes Form No.	New Form No. 527,479W	

Litho U.S.A.

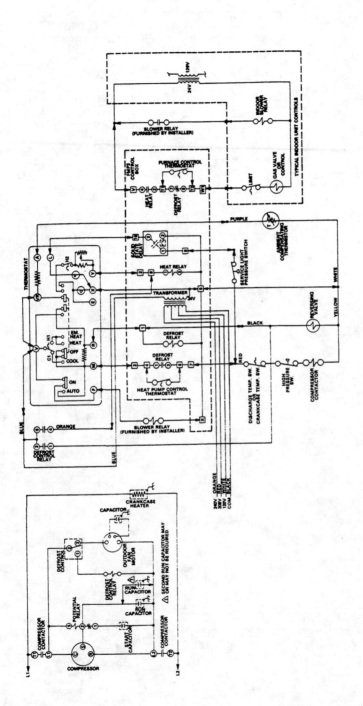

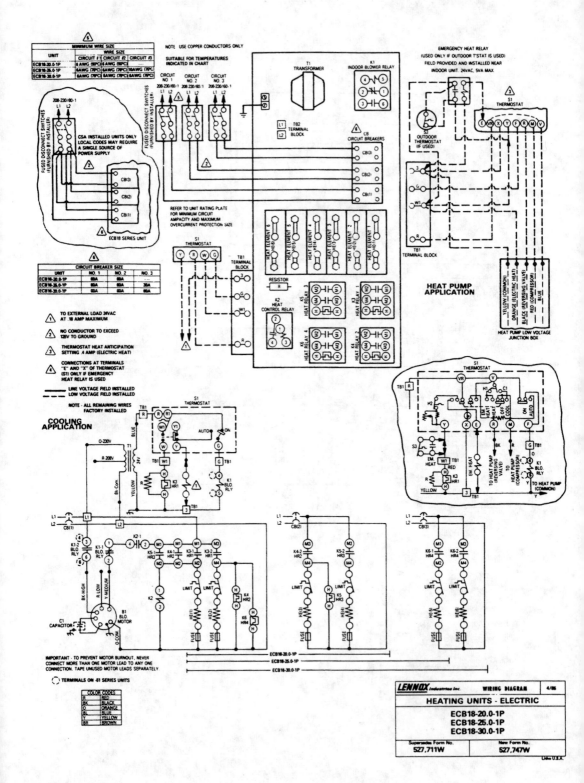

MINIMUM WIRE SIZE

UNIT	CIRCUIT #1	CIRCUIT #2	CIRCUIT #3
ECB18-20.0-1P	8 AWG (1PC)	6 AWG (1PC)	....
ECB18-25.0-1P	6AWG (1PC)	6AWG (1PC)	10AWG (1PC)
ECB18-30.0-1P	6AWG (1PC)	6AWG (1PC)	6AWG (1PC)

NOTE: USE COPPER CONDUCTORS ONLY

SUITABLE FOR TEMPERATURES INDICATED IN CHART

CIRCUIT BREAKER SIZE

UNIT	NO. 1	NO. 2	NO. 3
ECB18-20.0-1P	60A	60A	....
ECB18-25.0-1P	60A	60A	30A
ECB18-30.0-1P	60A	60A	60A

⚠ TO EXTERNAL LOAD 24VAC AT .18 AMP MAXIMUM

⚠ NO CONDUCTOR TO EXCEED 125V TO GROUND

⚠ THERMOSTAT HEAT ANTICIPATION SETTING .4 AMP (ELECTRIC HEAT)

⚠ CONNECTIONS AT TERMINALS "E" AND "X" OF THERMOSTAT (S1) ONLY IF EMERGENCY HEAT RELAY IS USED

—— LINE VOLTAGE FIELD INSTALLED
---- LOW VOLTAGE FIELD INSTALLED

NOTE: ALL REMAINING WIRES FACTORY INSTALLED

COOLING APPLICATION

HEAT PUMP APPLICATION

REFER TO UNIT RATING PLATE FOR MINIMUM CIRCUIT AMPACITY AND MAXIMUM OVERCURRENT PROTECTION SIZE

IMPORTANT - TO PREVENT MOTOR BURNOUT, NEVER CONNECT MORE THAN ONE MOTOR LEAD TO ANY ONE CONNECTION. TAPE UNUSED MOTOR LEADS SEPARATELY.

TERMINALS ON -61 SERIES UNITS

COLOR CODES

R	RED
BK	BLACK
O	ORANGE
BL	BLUE
Y	YELLOW
BR	BROWN

LENNOX Industries Inc. — **WIRING DIAGRAM** — 4/86

HEATING UNITS - ELECTRIC

ECB18-20.0-1P
ECB18-25.0-1P
ECB18-30.0-1P

Supersedes Form No.	New Form No.
527,711W	527,747W

222

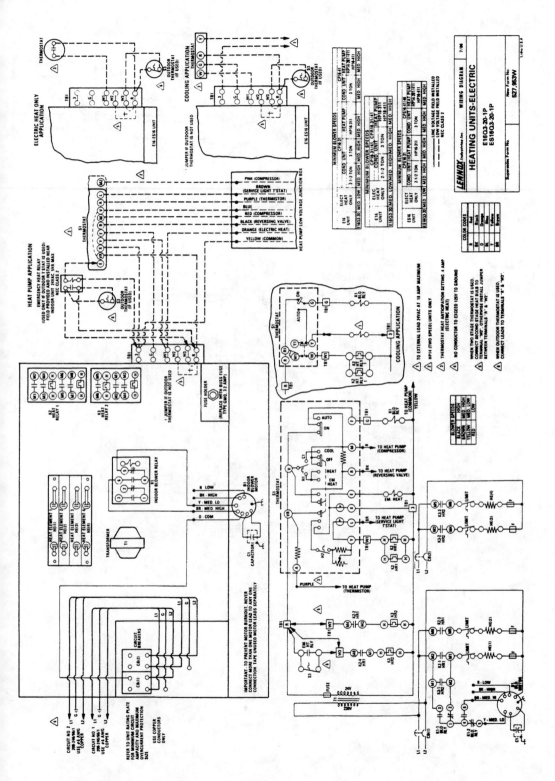

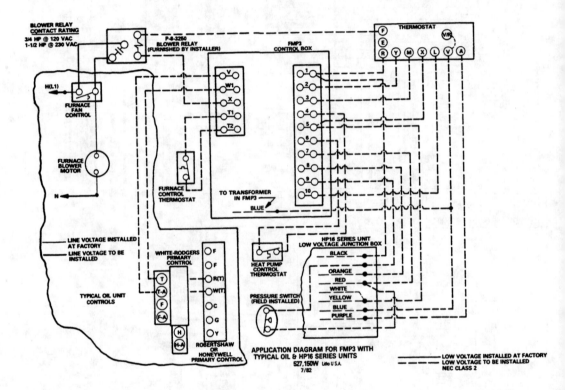

APPLICATION DIAGRAM FOR FMP3 WITH
TYPICAL OIL & HP16 SERIES UNITS
527,150W Litho U.S.A.
7/82

LENNOX WIRING DIAGRAM | 7/82

ACCESSORIES

Supersedes Form No.	New Form No.
	527,152W

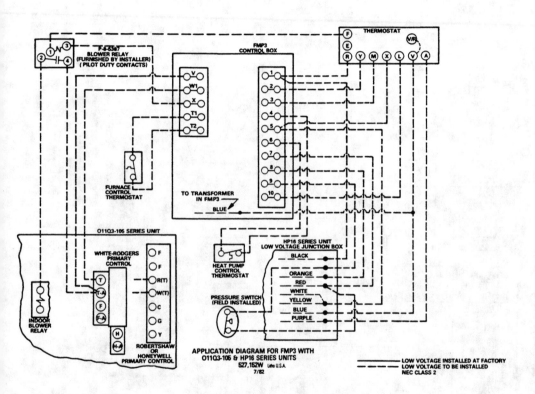

APPLICATION DIAGRAM FOR FMP3 WITH
O11Q3-105 & HP16 SERIES UNITS
527,152W Litho U.S.A.
7/82

LOW VOLTAGE INSTALLED AT FACTORY
LOW VOLTAGE TO BE INSTALLED
NEC CLASS 2

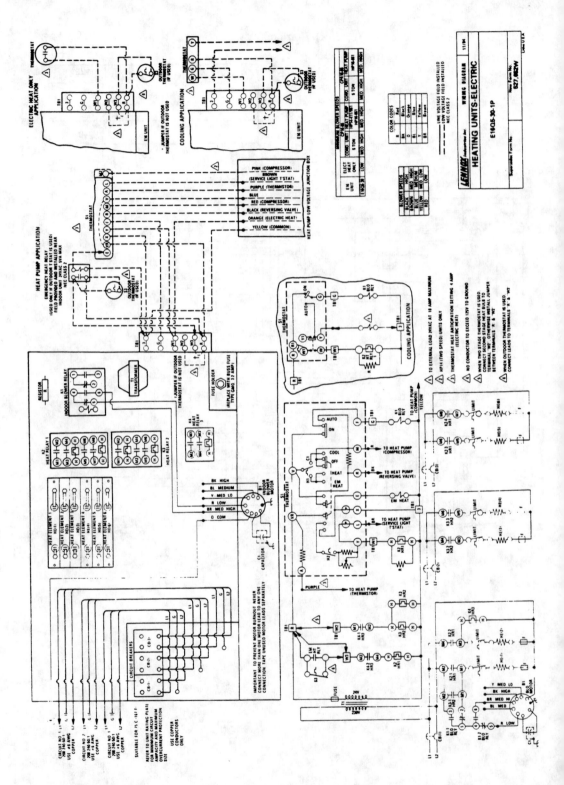

MAGIC CHEF WIRING DIAGRAMS

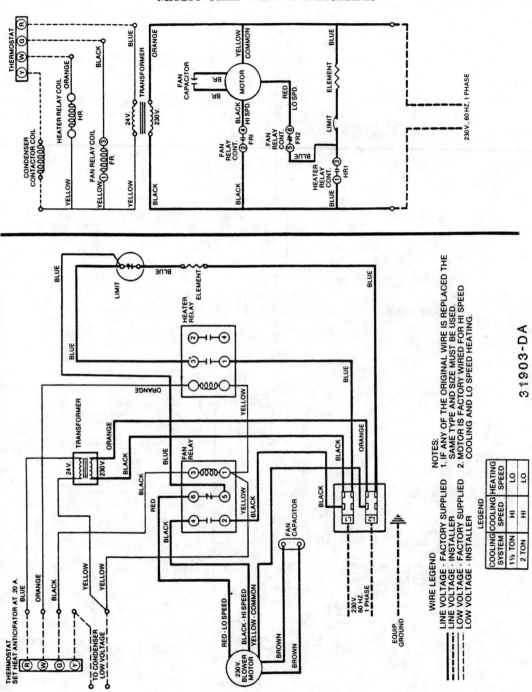

31903-DA

227

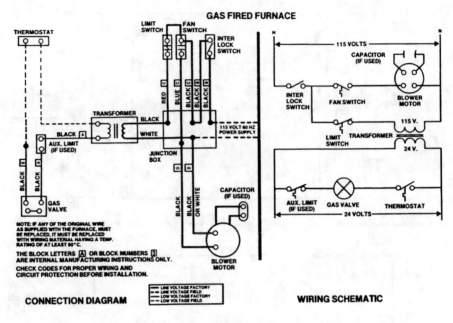

33210D2

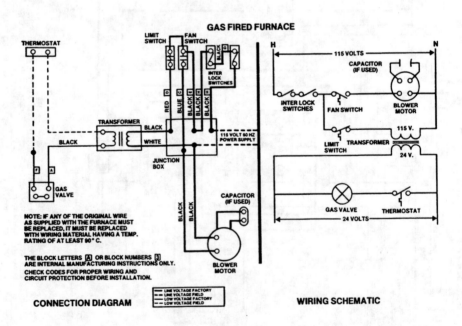

33218D2

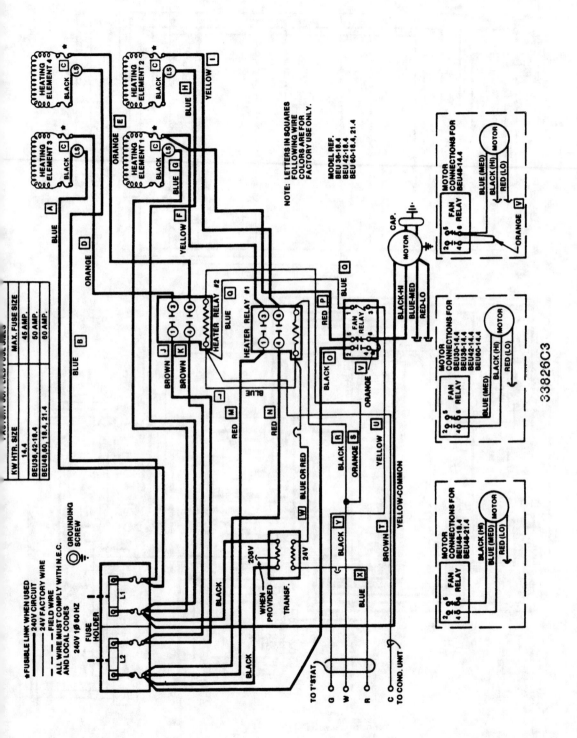

33826C3

229

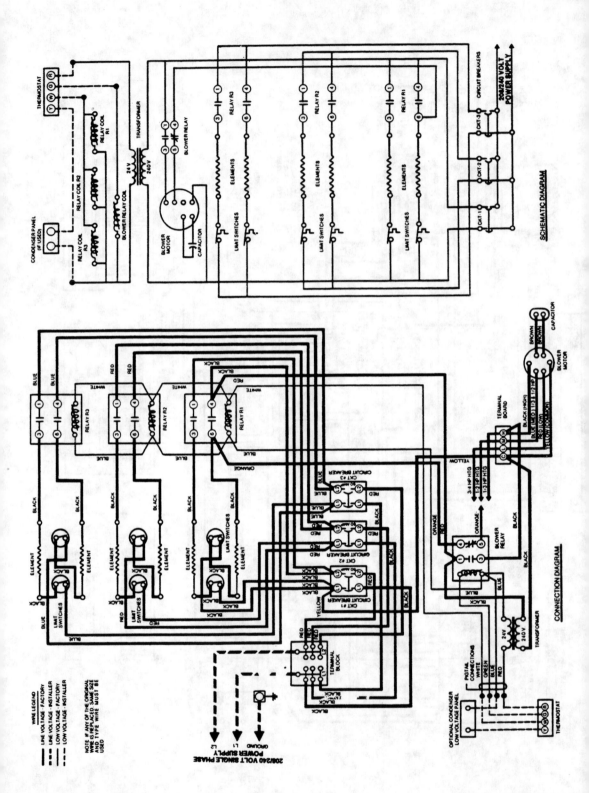

SCHEMATIC DIAGRAM

CONNECTION DIAGRAM

230

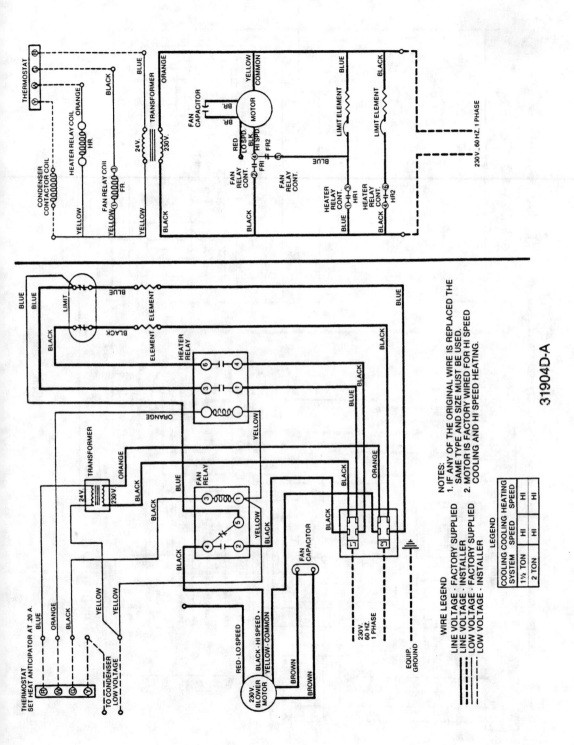

31904D-A

NOTES:
1. IF ANY OF THE ORIGINAL WIRE IS REPLACED THE SAME TYPE AND SIZE MUST BE USED.
2. MOTOR IS FACTORY WIRED FOR HI SPEED COOLING AND HI SPEED HEATING.

WIRE LEGEND
LINE VOLTAGE - FACTORY SUPPLIED
LINE VOLTAGE - INSTALLER
LOW VOLTAGE - FACTORY SUPPLIED
LOW VOLTAGE - INSTALLER

LEGEND

SYSTEM	COOLING SPEED	COOLING HEATING SPEED
1½ TON	HI	HI
2 TON	HI	HI

231

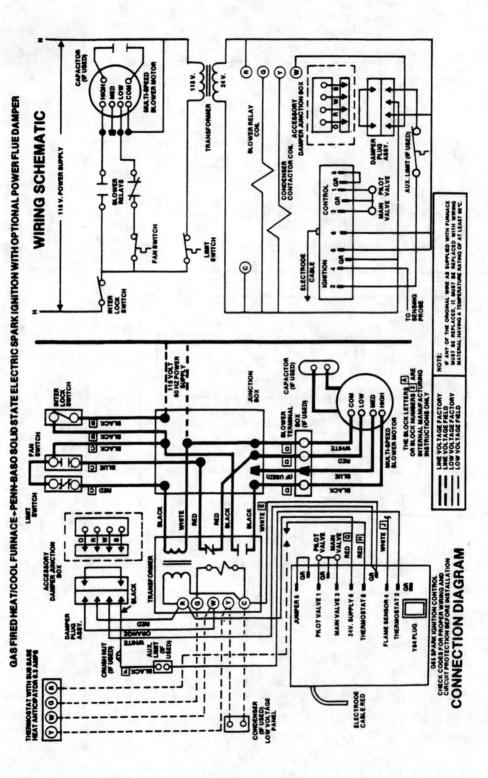

GAS FIRED HEAT/COOL FURNACE—PENN-BASO SOLID STATE ELECTRIC SPARK IGNITION WITH OPTIONAL POWER FLUE DAMPER

WIRING SCHEMATIC

CONNECTION DIAGRAM

33793D2

232

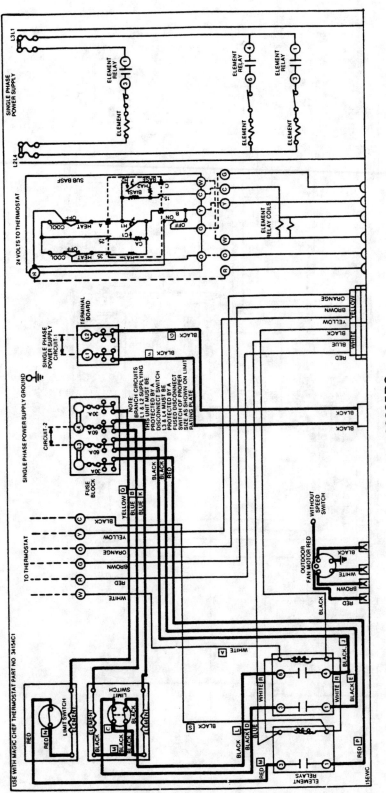

33985D2

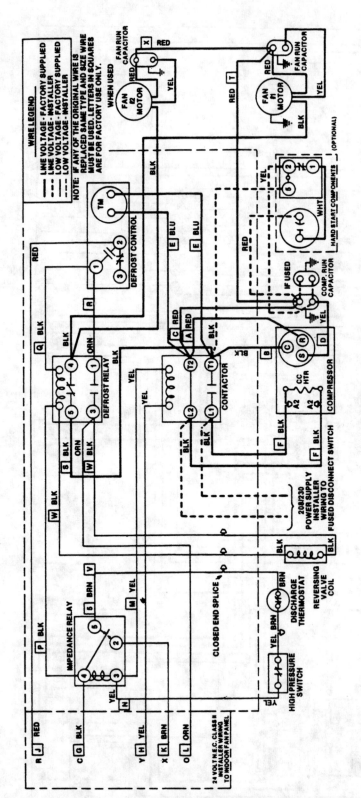

35479D3

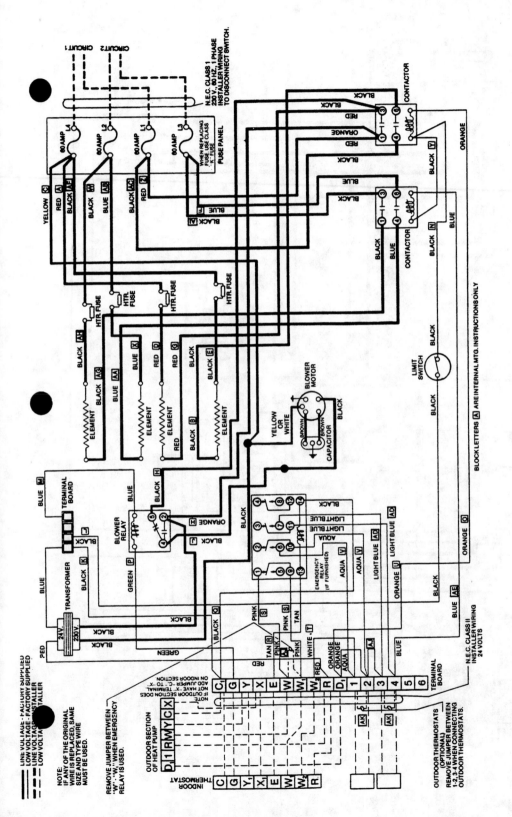

31553D4

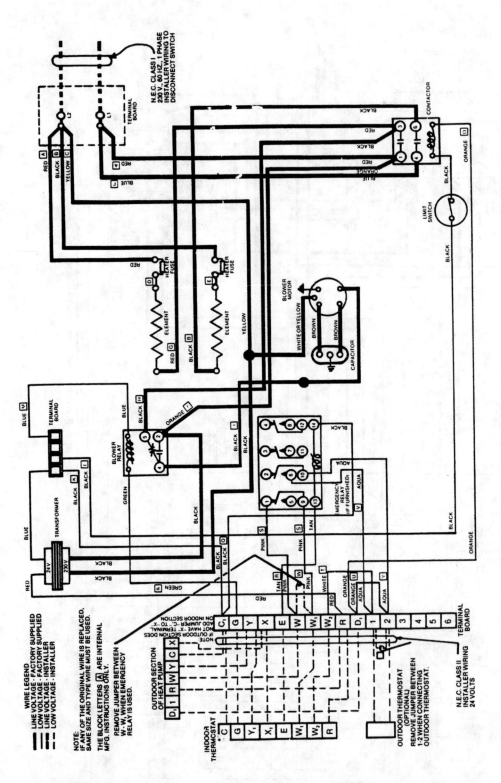

31594D3

236

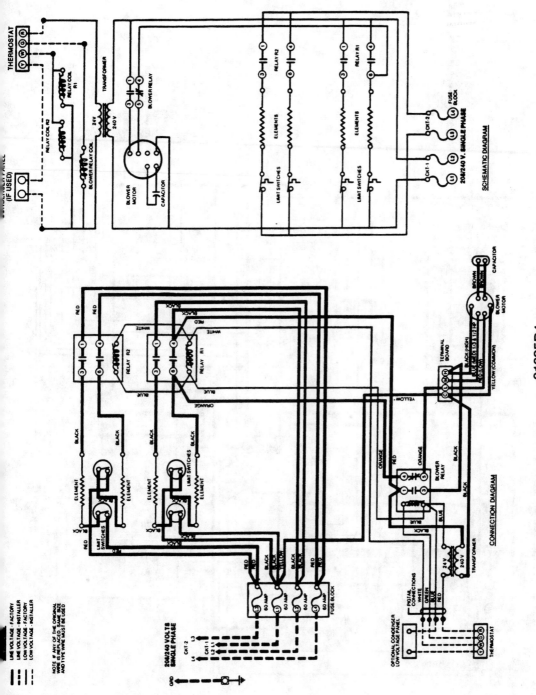

31395B4

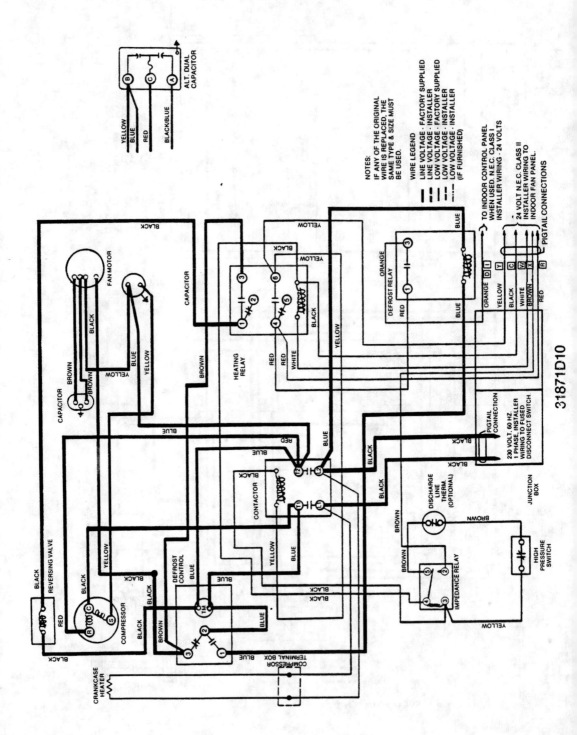

31871D10

238

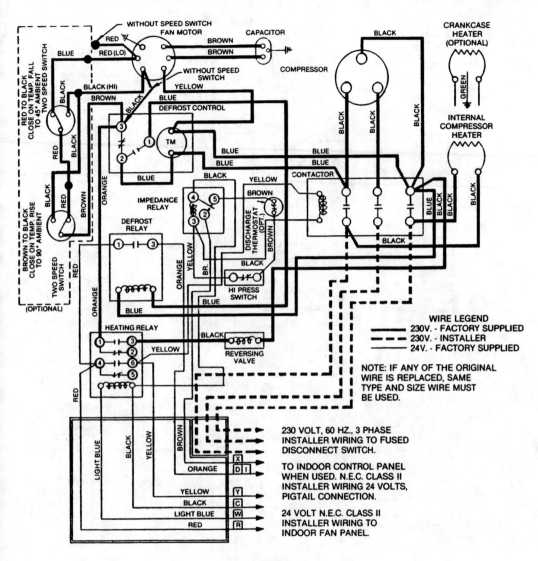

34646C1

GAS FIRED HEAT/COOL FURNACE

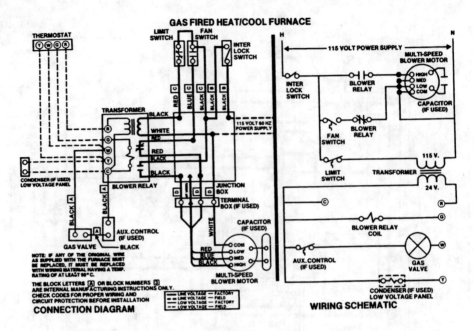

33215D3

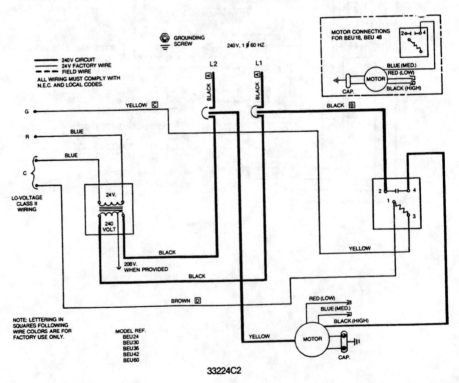

33224C2

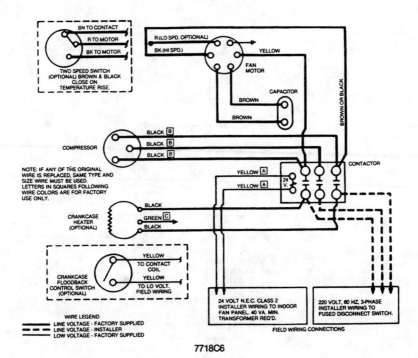

7718C6

ALL WIRING MUST COMPLY
WITH N.E.C. & LOCAL CODES.

240V CIR. WIRE
24V FAC. WIRE
FIELD WIRE

L1   L2

208/240V, 1 Ø, 60 HZ
CLASS I WIRING
MAX. SUPPLY VOLTAGE
TO GROUND 120V.

GROUNDING
SCREW

CIRCUIT
BREAKER
30 AMP

YELLOW
COMMON

240V

24V
TRANS.

BLUE OR RED

BLUE

BLACK C

LIMIT SWITCH

BLUE G

NOTE: LETTERING IN SQUARES
ARE USED FOR MANUFACTURING
PURPOSES ONLY.
IF ANY OF THE ORIGINAL WIRE IS
REPLACED, THE SAME TYPE & SIZE
MUST BE USED.

BLACK O

RED P

YEL F

HTR.
RELAY 1

RED M

FAN
RELAY

BROWN U

RED OR BLUE

ORANGE S

BLACK J

WHITE R

TRANSFORMER HEAT
ANTICIPATOR SETTING 0.20.

MOTOR

CAP

BLACK

34770B2

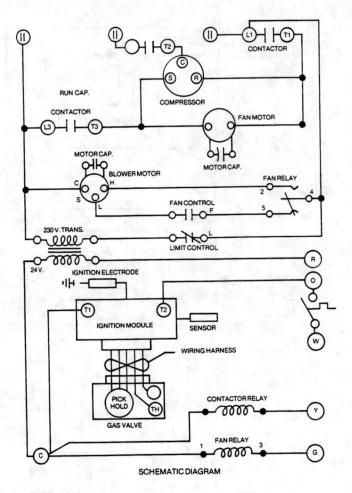

SCHEMATIC DIAGRAM

9-22026-9

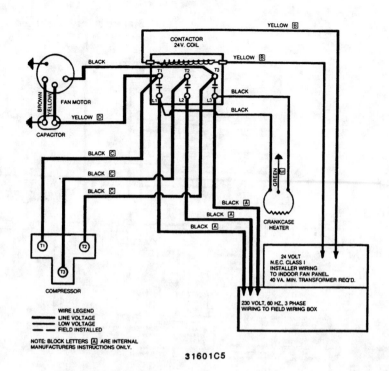

CONTACTOR
24 V. COIL

YELLOW B

YELLOW B

BLACK

FAN MOTOR

BLACK

BROWN

YELLOW

YELLOW D

CAPACITOR

BLACK C

BLACK C

BLACK C

BLACK A

BLACK A

BLACK A

GREEN E

CRANKCASE HEATER

T1   T2

T3

COMPRESSOR

24 VOLT
N.E.C. CLASS I
INSTALLER WIRING
TO INDOOR FAN PANEL.
40 VA. MIN. TRANSFORMER REQ'D.

230 VOLT, 60 HZ., 3 PHASE
WIRING TO FIELD WIRING BOX

WIRE LEGEND
LINE VOLTAGE
LOW VOLTAGE
FIELD INSTALLED

NOTE: BLOCK LETTERS A ARE INTERNAL
MANUFACTURERS INSTRUCTIONS ONLY.

31601C5

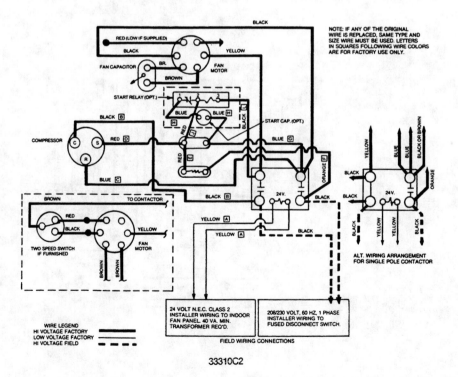

BLACK

RED (LOW IF SUPPLIED)

BLACK

YELLOW

FAN CAPACITOR

BR.

BROWN

FAN MOTOR

NOTE: IF ANY OF THE ORIGINAL
WIRE IS REPLACED, SAME TYPE AND
SIZE WIRE MUST BE USED. LETTERS
IN SQUARES FOLLOWING WIRE COLORS
ARE FOR FACTORY USE ONLY.

START RELAY (OPT.)

BLUE   BLUE H

H   RED

START CAP. (OPT.)

BLACK B

COMPRESSOR

C   S

R

RED D

RED

BLUE G

ORANGE F

YELLOW

BLUE

BLUE

BLACK OR BROWN

BLUE C

BLACK B

24 V.

BLACK

BLACK

ORANGE

BLACK

24 V.

BLACK

BROWN

RED

BLACK

YELLOW

YELLOW A

YELLOW A

BLACK

BLACK

YELLOW

YELLOW

TWO SPEED SWITCH
IF FURNISHED

TO CONTACTOR

FAN MOTOR

BROWN   BROWN

ALT. WIRING ARRANGEMENT
FOR SINGLE POLE CONTACTOR

24 VOLT N.E.C. CLASS 2
INSTALLER WIRING TO INDOOR
FAN PANEL. 40 VA. MIN.
TRANSFORMER REQ'D.

208/230 VOLT, 60 HZ., 1 PHASE
INSTALLER WIRING TO
FUSED DISCONNECT SWITCH.

FIELD WIRING CONNECTIONS

WIRE LEGEND
HI VOLTAGE FACTORY
LOW VOLTAGE FACTORY
HI VOLTAGE FIELD

33310C2

243

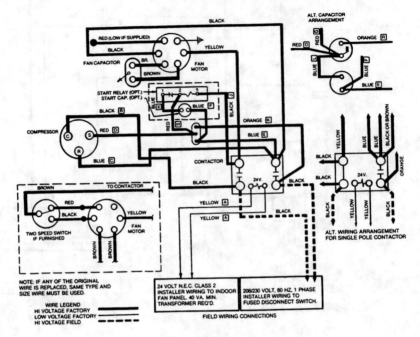

33150C3

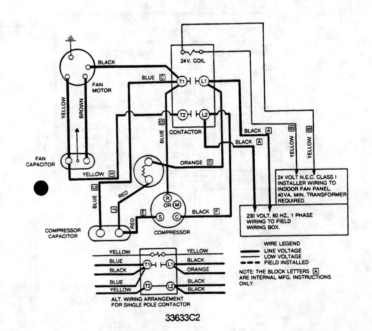

33633C2

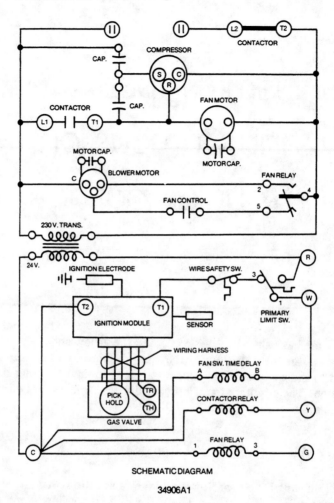

SCHEMATIC DIAGRAM

34906A1

5 Flame Types and Combustion Troubleshooting

There are basically two types of flames: the yellow and the blue. There are different variations of these two flames that are produced by changes in the primary air supplied to the gaseous fuels. Primary air is that air that is mixed with the gas before ignition. The counterpart of primary air is secondary air, which is mixed with the flame after ignition (see Fig. 5-1).

5-1 FLAME TYPES

The following is a list of flames and their indications that are encountered in heating systems using gas as a fuel. The first will be the yellow flame and we will progress to the blue flame.

Yellow Flame

This flame has a small blue-colored area at the bottom of the flame (see Fig. 5-2). The outer portion, or outer envelope, is completely yellow and is usually smoking. This smoke is unburned carbon from the fuel. The yellow, or luminous, portion of the flame is caused by the slow burning of the carbon that is being burned. Soot is the outcome of the yellow flame, which produces a lower-than-normal temperature than does the blue flame. A yellow flame is an indication of insufficient primary air. Incomplete combustion is also indicated by a yellow flame—a hazardous condition that cannot be allowed.

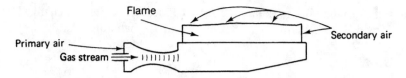

Figure 5-1.   Primary air and secondary air supporting combustion.

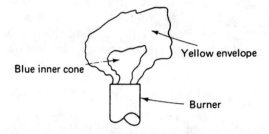

Figure 5-2.   Yellow flame.

Yellow-Tipped Flame

This condition occurs when not quite enough primary air is admitted to the burner. The yellow tips will be on the upper portion of the outer mantle (see Fig. 5-3). This flame is also undesirable because the unburned carbon will be deposited in the furnace flues and restrictors, thus eventually restricting the passage of the products of combustion.

Orange Color

Some orange or red color in a flame, usually in the form of streaks, should not cause any concern. These streaks are caused by dust particles in the air and cannot be completely eliminated (see Fig. 5-4). However, when there is a great amount of

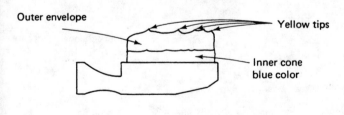

Figure 5-3.   Yellow-tipped flame.

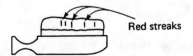

Figure 5-4.   Orange streaks in flame.

red or orange in the flame, the furnace location should be changed or some means provided for filtering the combustion air to the equipment.

Soft, Lazy Flame

This condition appears only when just enough primary air is admitted to the burner to cause the yellow flame tips to disappear. The inner cone and the outer envelope will not be as clearly defined as they are in the correct blue flame (see Fig. 5–5). This type of flame is best for high–low fire operation. It is not suitable, however, wherever there may be a shortage of secondary air. This flame will burn in the open air; however, when it touches a cooler surface, soot will be deposited on that surface.

Sharp, Blue Flame

When the proper ratio of gas to air is maintained, there will be a sharp, blue flame (see Fig. 5–6). Both the outer envelope and the inner cone will be pointed and the sides will be straight. The flame will be resting on the burner ports and there will not be any noticeable blowing noise. The flame will ignite smoothly on demand from the thermostat. Also, it will burn with a nonluminous flame. This is the most desirable flame for heating purposes on standard heating units.

Lifting Flame

When too much primary air is admitted to the burner, the flame will actually lift off the burner ports. This flame is undesirable for many reasons. When the flame is raised from the ports, there is a possibility that intermittent products of combustion will escape to the atmosphere (see Fig. 5–7). The flame will be small and will be accompanied by a blowing noise, much like that made by a blow torch. There will be rough ignition, and if enough secondary air is available, ignition may

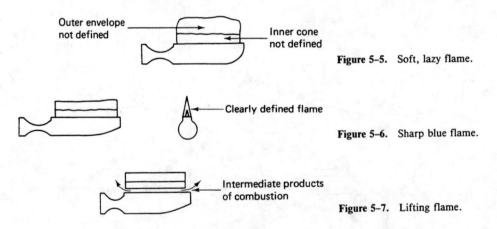

Figure 5–5.   Soft, lazy flame.

Figure 5–6.   Sharp blue flame.

Figure 5–7.   Lifting flame.

be impossible. Intermediate products of combustion are the same as the products of incomplete combustion, which must be avoided.

Floating Flame

A floating flame is an indication that there is a lack of secondary air to the flame. In severe cases, the flame will leave the burner and have the appearance of a floating cloud (see Fig. 5–8). Again, the intermediate products of combustion are apt to escape from between the burner and the flame. The flame is actually floating from place to place wherever sufficient secondary air for combustion may be found.

This flame is sometimes hard to detect because, as the door to the equipment room is opened, sufficient air will be admitted to allow the flame to rest on the burner properly. However, the obnoxious odor of aldehydes may still be detected. Another possible cause of this flame is an improperly operating vent system. If the products of combustion cannot escape, fresh air cannot reach the flame. This is a hazardous situation that must be eliminated regardless of the cause.

5–2 COMBUSTION TROUBLESHOOTING

Heating service technicians are called on to make fast, accurate analyses and corrections of heating equipment in order to provide the customer with safe, economical, and adequate heat. The major situations dealing with the flame that technicians will be called upon to analyze are (1) delayed ignition, (2) roll-out ignition, (3) flashback, (4) resonance, (5) yellow flame, (6) floating main burner flame lifting off the burners, and (7) main burner flame too large. In the following paragraphs we discuss these problems and make recommendations for their solution.

Delayed Ignition

Delayed ignition is caused by improper or poor flame travel to the main burner, or by poor flame distribution over the burner itself, or by too small orifices. When this situation occurs, it can usually be detected by a noisy ignition—that is, a light explosion, or puff, on ignition of the main burner.

The possible causes and correction of delayed ignition are:

1. Distorted burner or distorted carryover wing slots. Both the carryover wing slots and the main burner slots should be uniformly shaped and of the proper size

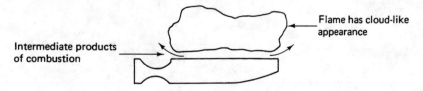

Figure 5–8. Floating flame.

(see Fig. 5–9). If not, the distortion could cause the adjacent burner to have faulty ignition accompanied by an accumulation of gas, which causes a noisy and dangerous ignition.

2. Misaligned carryover wings. The purpose of the carryover wing is to direct the gas for ignition to the main burner. The carryover wing on each burner must be aligned with the adjacent burner so that the flame path is no more than ⅟₁₆ in. above or ⅛ in. below the adjacent burner. A delayed ignition could result if these two limits are not met (see Fig. 5–10). This problem can be solved by loosening the clamp, aligning the main burner, and retightening the clamp.

3. Painted-over or rusted carryover wing slots. Occasionally, paint will get into the carryover wings during their manufacture. Also, the burners and heat exchanger will rust and this rust will accumulate in the ports and slots, preventing proper flame travel during the ignition period. This poor flame travel results in delayed ignition. The solution to this problem is to clean the ports and slots of paint and rust. If rust is the problem, the owner should be instructed to turn off the pilot during summer operation, to reduce the possibility of rust recurring if condensation forms on the colder surfaces of the furnace heat exchanger.

4. Low manifold gas pressure. If the supply gas pressure to the furnace should be too low, the flow of ignition gas may be slow from one burner to the next, possibly resulting in delayed ignition of some of the burners. This possibility exists on any installation, especially during cold weather, when the demand for gas is high. All furnaces are equipped with a pressure tap on the main gas valve or the manifold to measure the gas pressure. Use either a water manometer or a manifold gas pressure gauge (see Fig. 5–11). The manifold gas pressure for natural gas is 3½ to 4 in. of water column. Propane equipment requires an 11-in. water column pressure.

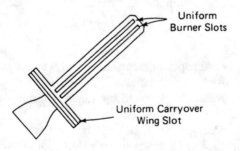

Figure 5–9. Carryover wing and burner slots.

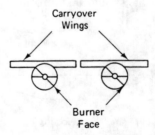

Figure 5–10. Carryover wing adjustment.

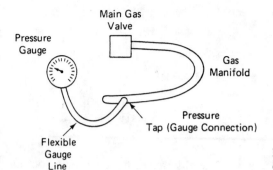

Figure 5-11. Manifold pressure connection.

These pressures should be taken with all other appliances on the same gas main in operation.

5. Orifices too small: The only solution to this problem is to install the proper-size orifices. The ones that are in the furnace may be increased in size, or new ones may be installed. Before changing the orifice size, make certain that the manifold gas pressure is properly adjusted. Then calculate the size needed and make the necessary correction. This condition is more likely to exist on new installations or units that have been changed from LP to natural gas.

6. Main gas valve equipped with step-opening regulator. When delayed ignition occurs on a furnace equipped with a step-opening gas valve, the remedies are: (a) to carefully check and correct any of the previously mentioned possibilities, and (b) next, to measure the first step opening of the gas pressure with a manometer or manifold pressure gauge. If the first step pressure opening has less than 2 in. of water column, there are two options that may be taken: (1) the first step pressure may be adjusted to at least 2 in. of water column; or (2) replace the step opening regulator with a standard regulator.

Roll-Out Ignition

Roll-out is usually indicated by a puff or "whish" sound on burner ignition. This situation is usually caused by two conditions: (a) soot or some other obstruction in the heat exchanger causing a restriction; or (b) a fast or quick opening gas valve which allows an accumulation of unburned gas in the combustion zone to ignite late, causing delayed ignition. The solution is to remove the soot or obstruction from the heat exchanger. In the case of a fast-opening gas valve, a surge arrestor may be installed in the bleed line from the regulator (see Fig. 5-12). The surge arrestor causes the rate of bleed-off to be reduced, resulting in a slower-opening gas valve and thus providing a smoother ignition and minimum roll-out on ignition.

Flashback

Flashback occurs when the flow of gas–air mixture entering the burner is slower than the velocity of the flame travel. This results in a popping sound and sometimes

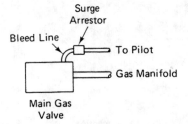

Figure 5–12. Surge arrestor location.

a flame burning at the orifice. It is usually caused by a low manifold gas pressure, a change in the gas–air mixture, or a sudden downdraft on the flame pattern. The following are possible causes of flashbacks:

1. Extremely hard flame
2. Distorted burner or carryover wing slots
3. Low manifold gas pressure
4. Defective burner orifice
5. Misaligned burner orifice
6. Erratic operation of the main gas valve
7. Unstable gas supply pressure
8. Rust or soot in the burner
9. Wrong type of LP gas, or gas–air supply mixture

Extremely hard flame.   In this situation, close off the primary air shutter until a small yellow tip appears on the flame. A yellow tip of up to 3 in. can be tolerated by most furnaces when they are hot. Then open the shutter until the yellow tips just disappear.

Distorted burner or carryover wing slots.   When this situation occurs, check the burner ports and carryover wing slots for damage. Correct any damage by repairing or replacing the burner.

Low manifold gas pressure.   To correct this condition, check the manifold pressure with a manometer or manifold pressure gauge after the burners have been fired for at least 5 minutes. Correct any pressures that do not meet the recommendations for the type of gas used.

Defective burner orifice.   To determine whether or not the burner orifice is bad, turn off the main gas valve, remove the burners, then remove each individual orifice from the manifold. Check the outside of the orifice spuds for dents or misshaping of the orifice hole. Check the inside of the orifices for foreign matter, such as dirt, grease, pipe dope, and so on. Remove any foreign material, being careful not to damage the orifice. Replace the orifices in the manifold and check for gas leaks with a soap-and-water solution or other liquid-type leak detector. Never use a match to leak-test gas piping.

Misaligned burner orifices.   If flashback is persistent on any burner, first check for damage and foreign matter. If the problem still exists, completely remove the manifold assembly from the furnace and apply water pressure to the assembly. This procedure will indicate any orifice misalignment by a crooked spray of water from a misaligned orifice (see Fig. 5-13). Make any necessary repairs to correct the problem. In some cases, repair may require the replacement of the entire assembly.

Unstable gas supply pressure.   This condition is more likely to occur on LP gas furnaces. The condition can be identified by use of a manifold gas pressure gauge or manometer connected to the gas line. This situation is usually caused by a chattering regulator or the main gas line exposed to extremely cold temperatures. It is sometimes necessary to install two-stage regulators to eliminate this problem.

Rust or soot in the burner.   When this situation occurs, the main burners must be removed and all foreign matter removed from inside them. The cause of rusting should be determined and corrected. Sometimes rusting is caused by condensation if the pilot burner is left on during summer operation. If so, the user should be instructed to turn off the pilot during the cooling season.

Sooting of a furnace may be caused by low gas pressure and/or an improper setting of the primary air adjustment. Either of these conditions is generally accompanied by a large yellow flame. Precautions must be taken to prevent sooting of a furnace heat exchanger.

Wrong type of LP gas or gas mixture.   The furnace nameplate must be checked to be certain that the furnace is designed for the type of gas being supplied to it. Mixtures of butane and propane gases used in furnaces not approved for liquefied petroleum could be the cause of poor burner operation. They could result in flashback, sooting, and burner damage because of overheating. A quick check for this situation is to check the relief valve setting. Tanks equipped with relief valve settings of 100 to 150 psi will accommodate gas mixtures. Relief valve settings of 200 to 250 psi are used on tanks equipped for pure propane.

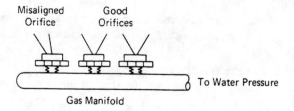

Figure 5-13.   Orifice misalignment.

Resonance

Resonance can be identified by a loud rumbling noise or a pure-tone buzz or hum. Resonance is most common when butane is used as the fuel. This situation can be caused by excessive primary air being supplied to the main burner or by a defective main burner spud. If excessive primary air is the cause, adjust the primary air shutter until a slight yellow tip appears on the flame. Lock the shutter in this position. If defective spuds are suspected, remove the spuds and check for nicks or dents on the outer edge of the orifice. Also, check for dirt or other foreign material inside the orifice. Clean or replace as necessary.

Yellow Flame

A large yellow, or luminous, flame is a good indication that almost all of the oxygen required for combustion is being taken from the secondary air. Complete combustion will occur as long as enough oxygen is supplied and nothing interferes with the flame. If the flame should touch a cooler surface, however, soot and toxic products will be released. Therefore, large yellow flames must be avoided.

There are five possible causes of large yellow flames:

1. Primary air shutter closed off too much; to correct, adjust the primary air shutter.
2. Partially clogged main burner ports or orifices; to correct, remove the spud and inspect for damage.
3. Misaligned burner spuds; to correct, check alignment and correct as necessary.
4. Soot or foreign material inside of the heat exchanger; to correct, clean the heat exchanger to allow proper combustion.
5. Poor venting, downdraft, or improper combustion air supply; to correct, eliminate any poor vent piping. It may sometimes be necessary to install an outside air duct to provide the required combustion air.

Floating Main Burner Flame

This situation is not very common, but may occur under the following conditions:

1. The heat exchanger is blocked with soot; to correct, clean the soot from the heat exchanger and correct the cause of poor combustion.
2. Air blowing into the heat exchanger; to correct, check for a leaking or cracked heat exchanger that may allow circulating air into the heat exchanger and correct.
3. Negative interior pressure in the furnace room; to correct, recheck the combustion air requirements. Also, check for exhaust fans in the furnace room. Correct as necessary.

Main Burner Flame Too Large

The possible causes and corrective action to be taken are:

1. Excessive gas manifold pressure; to correct, check the pressure with a manifold pressure gauge or manometer and reduce to the proper pressure.
2. Defective gas pressure regulator; to correct, replace the gas valve or pressure regulator.
3. Orifice size too large; to correct, replace the orifice with one of the proper size. Do not solder orifices closed and redrill, because the solder may melt out and cause more problems.

6 Safety Procedures

6-1 INTRODUCTION

The purpose of this list of safety procedures is to provide general safety precautions to be used by persons who own, install, operate, or maintain air-conditioning and refrigeration equipment. This list should not be used to replace instructions that are provided by the equipment manufacturer. Therefore, anyone attempting to work on air-conditioning and refrigeration equipment should be thoroughly familiar with the specific instructions for that particular unit.

The following is a list of the criteria that have been used here to indicate the intensity of the hazard:

Danger: This means that there is an immediate hazard that will result in severe personal injury or death.

Warning: This means that hazards or unsafe practices could result in severe personal injury or death.

Caution: This means that potential hazards or unsafe practices could result in minor personal injury.

Safety Instructions: These are general instructions that are necessary for safe working practices.

6-2 PERSONAL PROTECTION

Warning

1. Do not touch electrical wiring connections with wet hands.
2. Do not touch electrical equipment while standing on a wet surface or wearing wet shoes.
3. Do wear a hard hat or other head protection when there is a possibility of falling objects.

Caution

1. Do wear safety glasses equipped with side shields when working in manufacturing plants or at construction sites.
2. Do wear gloves when handling system components after a compressor motor burnout. The refrigerant and oil contain acid that can result in acid burns to the skin.
3. Do wear goggles and gloves when handling chemicals; when welding, cutting, grinding, or brazing; or when in an area where these operations are performed.
4. Do wear gloves and other protective clothing when working with sheet metal.
5. Do wear safety shoes when working around or lifting heavy objects.
6. Do wear protective clothing when arc welding to protect from serious burns.
7. Do wear hearing protection when working in areas where sound levels are greater than 90 decibels (dB).
8. Do not wear rings, jewelry, loose clothing, long ties, or gloves while working around moving belts and machinery.
9. Do not wear rings or watch while working on electrical equipment.

Safety Instructions

1. Keep your work area clean of debris and free of liquid spills on the floor.
2. Do not continue working if you become ill. An ill person is less observant and is therefore more subject to accidents.

6-3 RIGGING (USE OF CRANES)

Danger

Never use cranes under power lines. They may come in contact with the lines, causing a high-voltage electrical short.

Warning

1. Do check for the center of gravity before hoisting heavy equipment.
2. Do check for any specific hoisting instructions by the manufacturer before hoisting equipment.
3. Do check the component and assembly weights before assembling equipment to make certain the crane can lift the unit safely.
4. Do use only approved methods and rigging equipment.
5. Do not use eyebolt holes to hoist an entire assembly.
6. Do not move a loaded hoist, crane, or chain fall until the path is clear.
7. Do not use faulty rigging equipment.

Caution

1. Do not use rigging equipment when there is a possibility of slipping or losing your balance.
2. Do use platforms or catwalks to cross over a machine. Never climb over a machine.
3. Do not use ladders that are too straight up or have too much slope from top to bottom.

Safety Instructions

1. Do use lifting lugs according to hoisting instructions.
2. Do keep aware of where fellow workmen are at all times when hoisting equipment.
3. Do post signs indicating that heavy objects are being hoisted.

6-4 STORING AND HANDLING REFRIGERANT CYLINDERS

Warning

1. Do not heat a refrigerant cylinder with an open flame. When necessary to heat cylinder, use warm water.
2. Do not store refrigerant cylinders in direct sunlight.
3. Do not store refrigerant cylinders where the surrounding temperature may exceed the relief valve setting.
4. Do not reuse disposable refrigerant cylinders. It is dangerous and illegal.
5. Do not attempt to burn or incinerate refrigerant cylinders.
6. Do not alter the safety devices on a refrigerant cylinder.

7. Do not force connections.
8. Do not overfill refillable refrigerant cylinders.
9. Do open cylinder valves slowly to prevent rapid overpressurizing of the system.
10. Do use a proper wrench when opening and closing refrigerant cylinder valves.

Caution

1. Do not alter refrigerant cylinders.
2. Do not drop, dent, or abuse refrigerant cylinders.
3. Do not charge a cylinder with a refrigerant different from the color coding.
4. Do not entirely depend on the color of a refrigerant cylinder for refrigerant identification.
5. Do not overcharge a rechargeable refrigerant cylinder.
6. Do replace cylinder caps when cylinders are not in use.
7. Do avoid pressure surges when transferring refrigerant from one cylinder to another.
8. Do periodically inspect all hoses, fittings, and charging manifolds and replace them when needed.
9. Do secure all refrigerant cylinders to prevent damage. The following procedure is recommended:
   (a) *Large cylinder:* Lay on its side and prevent rolling by using chocks.
   (b) *Small cylinder:* Store in an upright position and secure with a strap or chain.

6-5 LEAK TESTING AND PRESSURE TESTING SYSTEMS

Danger

1. Do not use oxygen for pressurizing a refrigeration system. Oxygen and oil combine to cause an explosion.
2. Do not use full cylinder pressure when pressurizing a system with nitrogen.
3. Do not exceed the specified system test pressures when pressurizing a system.
4. Do not pressurize a system with nitrogen before putting in the refrigerant when leak testing. The nitrogen may have a greater pressure than the refrigerant cylinder can withstand.
5. Do use nitrogen when pressurizing a system above refrigerant pressures.
6. Do use a gauge-equipped regulator when pressurizing a system with nitrogen.
7. Do disconnect the nitrogen cylinder from the system when the system is pressurized.

6-6 REFRIGERANTS

Warning

1. Do not enter an enclosed area after a refrigerant leak has occurred without thoroughly ventilating the space. Use the buddy system or an approved air pack or both.
2. Do not allow liquid refrigerant to come into contact with the skin or eyes. Immediately wash the skin with soap and water. Immediately flush the eyes with water and consult a physician.
3. Do not breathe the fumes given off by a leak detector or open flame. The fumes are likely to be phosgene, a deadly gas.
4. Do use safety goggles.
5. Do use gloves when working with liquid refrigerants.

Caution

1. Do not weld or cut a refrigerant line or vessel until all the refrigerant has been removed.
2. Do not use an open flame in a space containing refrigerant vapor. Properly ventilate the area before entering.
3. Do not smoke in a space filled with refrigerant vapor.

Safety Instructions

1. Avoid heavy concentrations of refrigerant within an enclosed area. A refrigerant can displace enough oxygen to cause suffocation.
2. Do not allow heating devices, such as gas flames or electric elements, to operate in an area filled with refrigerant vapor. Heat can decompose the refrigerant into hazardous substances such as hydrochloric acid, hydrofluoric acid, and phosgene gas.
3. If a strong, irritating odor is detected, warn other persons and immediately leave the area. Report the problem to the proper persons.

6-7 RECIPROCATING COMPRESSORS

Danger

1. Do not work on electrical wiring until all electrical power is off.
2. Do not take ohmmeter measurements with the electrical power on.

Warning

1. Do not use a hermetic compressor for a vacuum pump. The winding can short out or a terminal may blow out, causing serious injury.
2. Do not use a welding torch when removing a compressor from the refrigerant system. The oil could catch fire and cause severe burns.
3. Do not purge refrigerant from the system through a cut, loosened connection, or broken pipe. Use the gauge manifold so that the rate of purging can be controlled.
4. Do not apply voltage to a compressor motor while the terminal box cover is off.
5. Do not attempt to loosen or remove bolts from the compressor while it is under pressure. Pump the system down to 0 to 2 psig.
6. Do not attempt to operate a compressor with the suction and discharge service valves closed.
7. Do open, tag, and lock all electrical disconnect switches while servicing the electrical circuits and connections.

Caution

Do valve off all the compressors in a multiple-compressor system before attempting to service any circuit. Otherwise, the oil equalizer connection will prevent purging pressure from a single compressor.

6-8 AIR HANDLING EQUIPMENT

Danger

1. Do not enter an enclosed fan cabinet while the unit is running.
2. Do not reach into a unit or fan cabinet while the unit is running.
3. Do not work on a fan or fan motor until the electrical disconnect switch is open, tagged, locked, and the fuses removed.
4. Do not work on electric circuits, heating elements, or connections until the electrical disconnect switch is open, tagged, locked, and the fuses removed.

Warning

1. Do not operate belt-driven equipment without the belt guards in place.
2. Do not service air control dampers until the operators are disconnected.
3. Do not handle access covers in high winds without sufficient help.
4. Do not pressurize a coil with a liquid refrigerant for leak testing.

5. Do not steam-clean coils until all personnel are clear of the area.

6. Do insure there is adequate ventilation when welding or cutting inside an air handling unit.

7. Do be sure that rooftop units are properly grounded.

Caution

1. Do not work on fans without securing the sheave with a rope or strap to prevent fan freewheeling.

2. Do not exceed the specified test pressure when pressurizing a system.

3. Do protect all flammable material when welding or cutting inside an air handling unit.

Safety Instructions

When operating and servicing air handling equipment, use good judgment and safe working practices to prevent damage to equipment and personal injury or property damage.

6-9 OXYACETYLENE WELDING AND CUTTING

Danger

1. Do not use oxygen for any purpose except welding and cutting.

2. Do not use oxygen to pressurize a refrigerant system.

Warning

1. Do not store oxygen cylinders near oil and/or grease.

2. Do not store oxygen cylinders near a combustible material.

3. Do not use oily hands or gloves to handle oxygen cylinders.

4. Do not weld or cut in an atmosphere filled with refrigerant vapor.

5. Do not weld or cut near combustible materials.

6. Do not weld or cut lines or pressure vessels until they have been properly evacuated.

7. Do not weld or cut unless adequate ventilation is provided.

8. Do use an approved breathing system and a buddy when necessary to weld or cut in an unventilated area.

9. Do wear goggles and welding gloves when welding or cutting.

Caution

1. Do not store oxygen and acetylene cylinders next to each other.
2. Do not store oxygen and acetylene cylinders near a heat source.
3. Do not block passageways, stairways, or ladders with welding equipment.
4. Do store oxygen and acetylene cylinders strapped or chained in an upright position.
5. Do wear suitable protective clothing when welding or cutting.

Safety Instructions

1. Do not use damaged or worn hoses.
2. Do not use connectors other than those made specifically for oxyacetylene welding and cutting equipment.
3. Do not stand in front of the regulator when opening a cylinder valve.
4. Do not ignite torches with anything other than a friction lighter.
5. Do observe the color coding of pipelines, cylinders, and hoses.
6. Do crack the cylinder valve before attaching the pressure regulator.
7. Do release tension on the regulator adjusting screw before opening the cylinder valve.
8. Do inspect the equipment for leaking shut-off valves and connections before use.

6-10 REFRIGERATION AND AIR-CONDITIONING MACHINERY (GENERAL)

Warning

1. Do not attempt to siphon refrigerants or other chemicals by mouth.
2. Do not attempt to weld, cut, or remove fittings while the system is pressurized.

Caution

1. Do not attempt to bend or step on pressurized refrigerant lines.
2. Do not weld or cut in an area containing refrigerant.
3. Do not loosen a packing gland nut before making sure there are plenty of threads engaged to prevent a blowout.
4. Do use replacement parts that meet the requirements of the equipment.
5. Do tag and valve off refrigerant, water, and steam lines before opening them.

6. Do periodically inspect fittings, piping, and valves for corrosion, leaks, damage, or rust.

Safety Instructions

1. Do be sure that all shipping bolts and plugs have been removed before initial startup.
2. Do periodically check all refrigerant and oil sight glasses for cracks.
3. Do use an ice-melting spray to remove ice from sight glasses.
4. Do not chip ice from sight glasses.

6-11 CENTRIFUGAL LIQUID CHILLERS (Heat Exchangers)

Danger

1. Do not exceed the specified test pressures when pressurizing a system.
2. Do not use oxygen to pressurize, purge, or leak-test a refrigerant system.
3. Do not make any pressure-relieving device inoperative.
4. Do not operate any machine without the proper pressure relief devices installed and functioning.

7 Useful Engineering Data*

*Courtesy of Copeland Refrigeration Corp.

FAHRENHEIT–CELSIUS TEMPERATURE CONVERSION CHART

The numbers in bold-face type in the center column refer to the temperature, either in Centigrade or Fahrenheit, which is to be converted to the other scale. If converting Fahrenheit to Centigrade, the equivalent temperature will be found in the left column. If converting Centigrade to Fahrenheit, the equivalent temperature will be found in the column on the right.

Cent.	C or F	Fahr	Cent.	C or F	Fahr	Cent.	C or F	Fahr	Cent.	C or F	Fahr
−40.0	−40	−40.0	−6.7	+20	+68.0	+26.7	+80	+176.0	+60.0	+140	+284.0
−39.4	−39	−38.2	−6.1	+21	+69.8	+27.2	+81	+177.8	+60.6	+141	+285.8
−38.9	−38	−36.4	−5.5	+22	+71.6	+27.8	+82	+179.6	+61.1	+142	+287.6
−38.3	−37	−34.6	−5.0	+23	+73.4	+28.3	+83	+181.4	+61.7	+143	+289.4
−37.8	−36	−32.8	−4.4	+24	+75.2	+28.9	+84	+183.2	+62.2	+144	+291.2
−37.2	−35	−31.0	−3.9	+25	+77.0	+29.4	+85	+185.0	+62.8	+145	+293.0
−36.7	−34	−29.2	−3.3	+26	+78.8	+30.0	+86	+186.8	+63.3	+146	+294.8
−36.1	−33	−27.4	−2.8	+27	+80.6	+30.6	+87	+188.6	+63.9	+147	+296.6
−35.6	−32	−25.6	−2.2	+28	+82.4	+31.1	+88	+190.4	+64.4	+148	+298.4
−35.0	−31	−23.8	−1.7	+29	+84.2	+31.7	+89	+192.2	+65.0	+149	+300.2
−34.4	−30	−22.0	−1.1	+30	+86.0	+32.2	+90	+194.0	+65.6	+150	+302.0
−33.9	−29	−20.2	−0.6	+31	+87.8	+32.8	+91	+195.8	+66.1	+151	+303.8
−33.3	−28	−18.4	.0	+32	+89.6	+33.3	+92	+197.6	+66.7	+152	+305.6
−32.8	−27	−16.6	+0.6	+33	+91.4	+33.9	+93	+199.4	+67.2	+153	+307.4
−32.2	−26	−14.8	+1.1	+34	+93.2	+34.4	+94	+201.2	+67.8	+154	+309.2
−31.7	−25	−13.0	+1.7	+35	+95.0	+35.0	+95	+203.0	+68.3	+155	+311.0
−31.1	−24	−11.2	+2.2	+36	+96.8	+35.6	+96	+204.8	+68.9	+156	+312.8
−30.6	−23	−9.4	+2.8	+37	+98.6	+36.1	+97	+206.6	+69.4	+157	+314.6
−30.0	−22	−7.6	+3.3	+38	+100.4	+36.7	+98	+208.4	+70.0	+158	+316.4
−29.4	−21	−5.8	+3.9	+39	+102.2	+37.2	+99	+210.2	+70.6	+159	+318.2
−28.9	−20	−4.0	+4.4	+40	+104.0	+37.8	+100	+212.0	+71.1	+160	+320.0
−28.3	−19	−2.2	+5.0	+41	+105.8	+38.3	+101	+213.8	+71.7	+161	+321.8
−27.8	−18	−0.4	+5.5	+42	+107.6	+38.9	+102	+215.6	+72.2	+162	+323.6
−27.2	−17	+1.4	+6.1	+43	+109.4	+39.4	+103	+217.4	+72.8	+163	+325.4
−26.7	−16	+3.2	+6.7	+44	+111.2	+40.0	+104	+219.2	+73.3	+164	+327.2
−26.1	−15	+5.0	+7.2	+45	+113.0	+40.6	+105	+221.0	+73.9	+165	+329.0
−25.6	−14	+6.8	+7.8	+46	+114.8	+41.1	+106	+222.8	+74.4	+166	+330.8
−25.0	−13	+8.6	+8.3	+47	+116.6	+41.7	+107	+224.6	+75.0	+167	+332.6
−24.4	−12	+10.4	+8.9	+48	+118.4	+42.2	+108	+226.4	+75.6	+168	+334.4
−23.9	−11	+12.2	+9.4	+49	+120.2	+42.8	+109	+228.2	+76.1	+169	+336.2
−23.3	−10	+14.0	+10.0	+50	+122.0	+43.3	+110	+230.0	+76.7	+170	+338.0
−22.8	−9	+15.8	+10.6	+51	+123.8	+43.9	+111	+231.8	+77.2	+171	+339.8
−22.2	−8	+17.6	+11.1	+52	+125.6	+44.4	+112	+233.6	+77.8	+172	+341.6
−21.7	−7	+19.4	+11.7	+53	+127.4	+45.0	+113	+235.4	+78.3	+173	+343.4
−21.1	−6	+21.2	+12.2	+54	+129.2	+45.6	+114	+237.2	+78.9	+174	+345.2
−20.6	−5	+23.0	+12.8	+55	+131.0	+46.1	+115	+239.0	+79.4	+175	+347.0
−20.0	−4	+24.8	+13.3	+56	+132.8	+46.7	+116	+240.8	+80.0	+176	+348.8
−19.4	−3	+26.6	+13.9	+57	+134.6	+47.2	+117	+242.6	+80.6	+177	+350.6
−18.9	−2	+28.4	+14.4	+58	+136.4	+47.8	+118	+244.4	+81.1	+178	+352.4
−18.3	−1	+30.2	+15.0	+59	+138.2	+48.3	+119	+246.2	+81.7	+179	+354.2
−17.8	0	+32.0	+15.6	+60	+140.0	+48.9	+120	+248.0	+82.2	+180	+356.0
−17.2	+1	+33.8	+16.1	+61	+141.8	+49.4	+121	+249.8	+82.8	+181	+357.8
−16.7	+2	+35.6	+16.7	+62	+143.6	+50.0	+122	+251.6	+83.3	+182	+359.6
−16.1	+3	+37.4	+17.2	+63	+145.4	+50.6	+123	+253.4	+83.9	+183	+361.4
−15.6	+4	+39.2	+17.8	+64	+147.2	+51.1	+124	+255.2	+84.4	+184	+363.2
−15.0	+5	+41.0	+18.3	+65	+149.0	+51.7	+125	+257.0	+85.0	+185	+365.0
−14.4	+6	+42.8	+18.9	+66	+150.8	+52.2	+126	+258.8	+85.6	+186	+366.8
−13.9	+7	+44.6	+19.4	+67	+152.6	+52.8	+127	+260.6	+86.1	+187	+368.6
−13.3	+8	+46.4	+20.0	+68	+154.4	+53.3	+128	+262.4	+86.7	+188	+370.4
−12.8	+9	+48.2	+20.6	+69	+156.2	+53.9	+129	+264.2	+87.2	+189	+372.2
−12.2	+10	+50.0	+21.1	+70	+158.0	+54.4	+130	+266.0	+87.8	+190	+374.0
−11.7	+11	+51.8	+21.7	+71	+159.8	+55.0	+131	+267.8	+88.3	+191	+375.8
−11.1	+12	+53.6	+22.2	+72	+161.6	+55.6	+132	+269.6	+88.9	+192	+377.6
−10.6	+13	+55.4	+22.8	+73	+163.4	+56.1	+133	+271.4	+89.4	+193	+379.4
−10.0	+14	+57.2	+23.3	+74	+165.2	+56.7	+134	+273.2	+90.0	+194	+381.2
−9.4	+15	+59.0	+23.9	+75	+167.0	+57.2	+135	+275.0	+90.6	+195	+383.0
−8.9	+16	+60.8	+24.4	+76	+168.8	+57.8	+136	+276.8	+91.1	+196	+384.8
−8.3	+17	+62.6	+25.0	+77	+170.6	+58.3	+137	+278.6	+91.7	+197	+386.6
−7.8	+18	+64.4	+25.6	+78	+172.4	+58.9	+138	+280.4	+92.2	+198	+388.4
−7.2	+19	+66.2	+26.1	+79	+174.2	+59.4	+139	+282.2	+92.8	+199	+390.2

From 1967 *ASHRAE Handbook of Fundamentals;* reprinted by permission.

PROPERTIES OF SATURATED STEAM: TEMPERATURE TABLE

Temp. F t	Abs. Press. Lb/Sq In. p	Specific Volume		Enthalpy			Entropy			Temp. F t
		Sat. Liquid v_f	Sat. Vapor v_g	Sat. Liquid h_f	Evap. h_{fg}	Sat. Vapor h_g	Sat. Liquid s_f	Evap. s_{fg}	Sat. Vapor s_g	
212	14.696	0.01672	26.80	180.07	970.3	1150.4	0.3120	1.4446	1.7566	212
214	15.289	0.01673	25.83	182.08	969.0	1151.1	0.3149	1.4385	1.7534	214
216	15.901	0.01674	24.90	184.10	967.8	1151.9	0.3179	1.4323	1.7502	216
218	16.533	0.01676	24.01	186.11	966.5	1152.6	0.3209	1.4262	1.7471	218
220	17.186	0.01677	23.15	188.13	965.2	1153.4	0.3239	1.4201	1.7440	220
222	17.861	0.01679	22.33	190.15	963.9	1154.1	0.3268	1.4141	1.7409	222
224	18.557	0.01680	21.55	192.17	962.6	1154.8	0.3298	1.4080	1.7378	224
226	19.275	0.01682	20.79	194.18	961.3	1155.5	0.3328	1.4020	1.7348	226
228	20.016	0.01683	20.07	196.20	960.1	1156.3	0.3357	1.3961	1.7318	228
230	20.780	0.01684	19.382	198.23	958.8	1157.0	0.3387	1.3901	1.7288	230
232	21.567	0.01686	18.720	200.25	957.4	1157.7	0.3416	1.3842	1.7258	232
234	22.379	0.01688	18.084	202.27	956.1	1158.4	0.3444	1.3784	1.7228	234
236	23.217	0.01689	17.473	204.29	954.8	1159.1	0.3473	1.3725	1.7199	236
238	24.080	0.01691	16.886	206.32	953.5	1159.8	0.3502	1.3667	1.7169	238
240	24.969	0.01692	16.323	208.34	952.2	1160.5	0.3531	1.3609	1.7140	240
242	25.884	0.01694	15.782	210.37	950.8	1161.2	0.3560	1.3551	1.7111	242
244	26.827	0.01696	15.262	212.39	949.5	1161.9	0.3589	1.3494	1.7083	244
246	27.798	0.01697	14.762	214.42	948.2	1162.6	0.3618	1.3436	1.7054	246
248	28.797	0.01699	14.282	216.45	946.8	1163.3	0.3647	1.3379	1.7026	248
250	29.825	0.01700	13.821	218.48	945.5	1164.0	0.3675	1.3323	1.6998	250
252	30.884	0.01702	13.377	220.51	944.2	1164.7	0.3704	1.3266	1.6970	252
254	31.973	0.01704	12.950	222.54	942.8	1165.3	0.3732	1.3210	1.6942	254
256	33.093	0.01705	12.539	224.58	941.4	1166.0	0.3761	1.3154	1.6915	256
258	34.245	0.01707	12.144	226.61	940.1	1166.7	0.3789	1.3099	1.6888	258
260	35.429	0.01709	11.763	228.64	938.7	1167.3	0.3817	1.3043	1.6860	260
262	36.646	0.01710	11.396	230.68	937.3	1168.0	0.3845	1.2988	1.6833	262
264	37.897	0.01712	11.043	232.72	936.0	1168.7	0.3874	1.2933	1.6807	264
266	39.182	0.01714	10.704	234.76	934.5	1169.3	0.3902	1.2878	1.6780	266
268	40.502	0.01715	10.376	236.80	933.2	1170.0	0.3930	1.2824	1.6753	268
270	41.858	0.01717	10.061	238.84	931.8	1170.6	0.3958	1.2769	1.6727	270
272	43.252	0.01719	9.756	240.88	930.3	1171.2	0.3986	1.2715	1.6701	272
274	44.682	0.01721	9.463	242.92	929.0	1171.9	0.4014	1.2661	1.6675	274
276	46.150	0.01722	9.181	244.96	927.5	1172.5	0.4041	1.2608	1.6649	276
278	47.657	0.01724	8.908	247.01	926.1	1173.1	0.4069	1.2554	1.6623	278
280	49.203	0.01726	8.645	249.06	924.7	1173.8	0.4096	1.2501	1.6597	280
282	50.790	0.01728	8.391	251.10	923.3	1174.4	0.4124	1.2448	1.6572	282
284	52.418	0.01730	8.146	253.15	921.8	1175.0	0.4152	1.2395	1.6547	284
286	54.088	0.01732	7.910	255.20	920.4	1175.6	0.4179	1.2343	1.6522	286
288	55.800	0.01733	7.682	257.26	918.9	1176.2	0.4207	1.2290	1.6497	288
290	57.556	0.01735	7.461	259.31	917.5	1176.8	0.4234	1.2238	1.6472	290
292	59.356	0.01737	7.248	261.36	916.0	1177.4	0.4261	1.2186	1.6447	292
294	61.201	0.01739	7.043	263.42	914.6	1178.0	0.4288	1.2134	1.6422	294
296	63.091	0.01741	6.844	265.48	913.1	1178.6	0.4315	1.2083	1.6398	296
298	65.028	0.01743	6.652	267.53	911.6	1179.1	0.4343	1.2031	1.6374	298
300	67.013	0.01745	6.466	269.59	910.1	1179.7	0.4369	1.1980	1.6350	300
310	77.68	0.01755	5.626	279.92	902.6	1182.5	0.4504	1.1727	1.6231	310
320	89.66	0.01765	4.914	290.28	894.9	1185.2	0.4637	1.1478	1.6115	320
330	103.06	0.01776	4.307	300.68	887.0	1187.7	0.4769	1.1233	1.6002	330
340	118.01	0.01787	3.788	311.13	879.0	1190.1	0.4900	1.0992	1.5891	340
350	134.63	0.01799	3.342	321.63	870.7	1192.3	0.5029	1.0754	1.5783	350
360	153.04	0.01811	2.957	332.18	862.2	1194.4	0.5158	1.0519	1.5677	360
370	173.37	0.01823	2.625	342.79	853.5	1196.3	0.5286	1.0287	1.5573	370
380	195.77	0.01836	2.335	353.45	844.6	1198.1	0.5413	1.0059	1.5471	380
390	220.37	0.01850	2.0836	364.17	835.4	1199.6	0.5539	0.9832	1.5371	390
400	247.31	0.01864	1.8633	374.97	826.0	1201.0	0.5664	0.9608	1.5272	400
410	276.75	0.01878	1.6700	385.83	816.3	1202.1	0.5788	0.9386	1.5174	410
420	308.83	0.01894	1.5000	396.77	806.3	1203.1	0.5912	0.9166	1.5078	420
430	343.72	0.01910	1.3499	407.79	796.0	1203.8	0.6035	0.8947	1.4982	430
440	381.59	0.01926	1.2171	418.90	785.4	1204.3	0.6158	0.8730	1.4887	440
450	422.6	0.0194	1.0993	430.1	774.5	1204.6	0.6280	0.8513	1.4793	450
460	466.9	0.0196	0.9944	441.4	763.2	1204.6	0.6402	0.8298	1.4700	460
470	514.7	0.0198	0.9009	452.8	751.5	1204.3	0.6523	0.8083	1.4606	470
480	566.1	0.0200	0.8172	464.4	739.4	1203.7	0.6645	0.7868	1.4513	480
490	621.4	0.0202	0.7423	476.0	726.8	1202.8	0.6766	0.7653	1.4419	490
500	680.8	0.0204	0.6749	487.8	713.9	1201.7	0.6887	0.7438	1.4325	500

From 1967 *ASHRAE Handbook of Fundamentals;* reprinted by permission.

DECIMAL EQUIVALENTS, AREAS, AND CIRCUMFERENCES OF CIRCLES

Diameter	Decimal Equivalent	Circumference	Area	Diameter	Decimal Equivalent	Circumference	Area	Diameter	Decimal Equivalent	Circumference	Area
1/64	.0156	.04909	.00019	3/4	.7500	2.356	.4418	3	3.000	9.425	7.069
1/32	.0312	.09817	.00077	49/64	.7656	2.405	.4604	3 1/16	3.0625	9.621	7.366
3/64	.0468	.1473	.00173	25/32	.7812	2.454	.4794	3 1/8	3.1250	9.817	7.670
				51/64	.7969	2.503	.4987	3 3/16	3.1875	10.01	7.980
1/16	.0625	.1963	.00307	13/16	.8125	2.553	.5185	3 1/4	3.2500	10.21	8.296
5/64	.0781	.2454	.00479	53/64	.8281	2.602	.5386	3 5/16	3.3125	10.41	8.618
3/32	.0937	.2945	.00690	27/32	.8437	2.651	.5591	3 3/8	3.3750	10.60	8.946
7/64	.1093	.3436	.00940	55/64	.8594	2.700	.5800	3 7/16	3.4375	10.80	9.281
1/8	.1250	.3927	.01227	7/8	.8750	2.749	.6013	3 1/2	3.5000	11.00	9.621
9/64	.1406	.4418	.01553	57/64	.8906	2.798	.6230	3 9/16	3.5625	11.19	9.968
5/32	.1562	.4909	.01917	29/32	.9062	2.847	.6450	3 5/8	3.6250	11.39	10.32
11/64	.1718	.5400	.02320	59/64	.9219	2.896	.6675	3 11/16	3.6875	11.58	10.68
3/16	.1875	.5890	.02761	15/16	.9375	2.945	.6903	3 3/4	3.7500	11.78	11.04
13/64	.2031	.6381	.03241	61/64	.9531	2.994	.7135	3 13/16	3.8125	11.98	11.42
7/32	.2187	.6872	.03758	31/32	.9687	3.043	.7371	3 7/8	3.8750	12.17	11.79
15/64	.2343	.7363	.04314	63/64	.9844	3.093	.7610	3 15/16	3.9375	12.37	12.18
1/4	.2500	.7854	.04909	1	1.000	3.142	.7854	4	4.000	12.57	12.57
17/64	.2656	.8345	.05542	1 1/16	1.0625	3.338	.8866	4 1/16	4.0625	12.76	12.96
9/32	.2812	.8836	.06213	1 1/8	1.1250	3.534	.9940	4 1/8	4.1250	12.96	13.36
19/64	.2968	.9327	.06920	1 3/16	1.1875	3.731	1.108	4 3/16	4.1875	13.16	13.77
5/16	.3125	.9817	.07670	1 1/4	1.2500	3.927	1.227	4 1/4	4.2500	13.35	14.19
21/64	.3281	1.031	.08456	1 5/16	1.3125	4.123	1.353	4 5/16	4.3125	13.55	14.61
11/32	.3437	1.080	.09281	1 3/8	1.3750	4.320	1.485	4 3/8	4.3750	13.74	15.03
23/64	.3593	1.129	.1014	1 7/16	1.4375	4.516	1.623	4 7/16	4.4375	13.94	15.47
3/8	.3750	1.178	.1104	1 1/2	1.5000	4.712	1.767	4 1/2	4.5000	14.14	15.90
25/64	.3906	1.227	.1198	1 9/16	1.5625	4.909	1.917	4 9/16	4.5625	14.33	16.35
13/32	.4062	1.276	.1296	1 5/8	1.6250	5.105	2.074	4 5/8	4.6250	14.53	16.80
27/64	.4218	1.325	.1398	1 11/16	1.6875	5.301	2.237	4 11/16	4.6875	14.73	17.26
7/16	.4375	1.374	.1503	1 3/4	1.7500	5.498	2.405	4 3/4	4.750	14.92	17.72
29/64	.4531	1.424	.1613	1 13/16	1.8125	5.694	2.580	4 13/16	4.8125	15.12	18.19
15/32	.4687	1.473	.1726	1 7/8	1.8750	5.890	2.761	4 7/8	4.8750	15.32	18.67
31/64	.4844	1.522	.1843	1 15/16	1.9375	6.087	2.948	4 15/16	4.9375	15.51	19.51
1/2	.5000	1.571	.1963	2	2.000	6.238	3.142	5	5.000	15.71	19.63
33/64	.5156	1.620	.2088	2 1/16	2.0625	6.480	3.341	5 1/16	5.0625	15.90	20.13
17/32	.5312	1.669	.2217	2 1/8	2.1250	6.676	3.547	5 1/8	5.1250	16.10	20.63
35/64	.5468	1.718	.2349	2 3/16	2.1875	6.872	3.758	5 3/16	5.1875	16.30	21.14
9/16	.5625	1.767	.2485	2 1/4	2.2500	7.069	3.976	5 1/4	5.2500	16.49	21.65
37/64	.5781	1.816	.2626	2 5/16	2.3125	7.265	4.200	5 5/16	5.3125	16.69	22.17
19/32	.5937	1.865	.2769	2 3/8	2.3750	7.461	4.430	5 3/8	5.3750	16.89	22.69
39/64	.6094	1.914	.2916	2 7/16	2.4375	7.658	4.666	5 7/16	5.4375	17.08	23.22
5/8	.6250	1.963	.3068	2 1/2	2.5000	7.854	4.909	5 1/2	5.5000	17.28	23.76
41/64	.6406	2.013	.3223	2 9/16	2.5625	8.050	5.157	5 9/16	5.5625	17.48	24.30
21/32	.6562	2.062	.3382	2 5/8	2.6250	8.247	5.412	5 5/8	5.6250	17.67	24.85
43/64	.6719	2.111	.3545	2 11/16	2.6875	8.443	5.673	5 11/16	5.6875	17.87	25.41
11/16	.6875	2.160	.3712	2 3/4	2.7500	8.639	5.940	5 3/4	5.7500	18.06	25.97
45/64	.7031	2.209	.3883	2 13/16	2.8125	8.836	6.213	5 13/16	5.8125	18.26	26.53
23/32	.7187	2.258	.4057	2 7/8	2.8750	9.032	6.492	5 7/8	5.8750	18.46	27.11
47/64	.7344	2.307	.4236	2 15/16	2.9375	9.228	6.777	5 15/16	5.9375	18.65	27.69

CONVERSION TABLE: INCHES INTO MILLIMETERS

Inches	Milli-meters	Inches	Milli-meters	Inches	Milli-meters	Inches	Milli-meters	Inches	Milli-meters	Inches	Milli-meters
1/64	0.3969	35/64	21.8281	2 13/32	61.1189	4 3/32	103.981	5 25/32	146.844	8 15/16	227.013
1/32	0.7937	7/8	22.2250	2 7/16	61.9126	4 1/8	104.775	5 13/16	147.638	9	228.600
3/64	1.1906	57/64	22.6219	2 15/32	62.7064	4 5/32	105.569	5 27/32	148.432	9 1/16	230.188
1/16	1.5875	29/32	23.0187	2 1/2	63.5001	4 3/16	106.363	5 7/8	149.225	9 1/8	231.775
5/64	1.9844	59/64	23.4156	2 17/32	64.2939	4 7/32	107.156	5 29/32	150.019	9 3/16	233.363
3/32	2.3812	15/16	23.8125	2 9/16	65.0876	4 1/4	107.950	5 15/16	150.813	9 1/4	234.950
7/64	2.7781	61/64	24.2094	2 19/32	65.8814	4 9/32	108.744	5 31/32	151.607	9 3/8	236.538
1/8	3.1750	31/32	24.6062	2 5/8	66.6751	4 5/16	109.538	6	152.400	9 3/8	238.125
9/64	3.5719	63/64	25.0031	2 21/32	67.4689	4 11/32	110.331	6 1/16	153.988	9 7/16	239.713
5/32	3.9687	1	25.4001	2 11/16	68.2626	4 3/8	111.125	6 1/8	155.575	9 1/2	241.300
11/64	4.3656	1 1/32	26.1938	2 23/32	69.0564	4 13/32	111.919	6 3/16	157.163	9 5/8	242.888
3/16	4.7625	1 1/16	26.9876	2 3/4	69.8501	4 7/32	112.713	6 1/4	158.750	9 5/8	244.475
13/64	5.1594	1 3/32	27.7813	2 25/32	70.6439	4 15/32	113.506	6 5/16	160.338	9 11/16	246.063
7/32	5.5562	1 1/8	28.5751	2 13/16	71.4376	4 1/2	114.300	6 3/8	161.925	9 3/4	247.650
15/64	5.9531	1 5/32	29.3688	2 27/32	72.2314	4 17/32	115.094	6 7/16	163.513	9 13/16	249.238
1/4	6.3500	1 3/16	30.1626	2 7/8	73.0251	4 9/16	115.888	6 1/2	165.100	9 7/8	250.825
17/64	6.7469	1 7/32	30.9563	2 29/32	73.8189	4 19/32	116.681	6 9/16	166.688	9 15/16	252.413
9/32	7.1437	1 1/4	31.7501	2 15/16	74.6126	4 5/8	117.475	6 5/8	168.275	10	254.001
19/64	7.5406	1 9/32	32.5438	2 31/32	75.4064	4 21/32	118.269	6 11/16	169.863	10 1/16	255.588
5/16	7.9375	1 5/16	33.3376	3	76.2002	4 11/16	119.063	6 3/4	171.450	10 1/8	257.176
21/64	8.3344	1 11/32	34.1313	3 1/32	76.9939	4 23/32	119.856	6 13/16	173.038	10 3/16	258.763
11/32	8.7312	1 3/8	34.9251	3 1/16	77.7877	4 3/4	120.650	6 7/8	174.625	10 1/4	260.351
23/64	9.1281	1 13/32	35.7188	3 3/32	78.5814	4 25/32	121.444	6 15/16	176.213	10 5/16	261.938
3/8	9.5250	1 7/16	36.5126	3 1/8	79.3752	4 13/16	122.238	7	177.800	10 3/8	263.526
25/64	9.9219	1 15/32	37.3063	3 5/32	80.1689	4 27/32	123.031	7 1/16	179.388	10 7/16	265.113
13/32	10.3187	1 1/2	38.1001	3 3/16	80.9627	4 7/8	123.825	7 1/8	180.975	10 1/2	266.701
27/64	10.7156	1 17/32	38.8938	3 7/32	81.7564	4 29/32	124.619	7 3/16	182.563	10 9/16	268.288
7/16	11.1125	1 9/16	39.6876	3 1/4	82.5502	4 15/16	125.413	7 1/4	184.150	10 5/8	269.876
29/64	11.5094	1 19/32	40.4813	3 9/32	83.3439	4 31/32	126.206	7 5/16	185.738	10 11/16	271.463
15/32	11.9062	1 5/8	41.2751	3 5/16	84.1377	5	127.000	7 3/8	187.325	10 3/4	273.051
31/64	12.3031	1 21/32	42.0688	3 11/32	84.9314	5 1/32	127.794	7 7/16	188.913	10 13/16	274.638
1/2	12.7000	1 11/16	42.8626	3 3/8	85.7252	5 1/16	128.588	7 1/2	190.500	10 7/8	276.226
33/64	13.0969	1 23/32	43.6563	3 13/32	86.5189	5 3/32	129.382	7 9/16	192.088	10 15/16	277.813
17/32	13.4937	1 3/4	44.4501	3 7/16	87.3127	5 1/8	130.175	7 5/8	193.675	11	279.401
35/64	13.8906	1 25/32	45.2438	3 15/32	88.1064	5 5/32	130.969	7 11/16	195.263	11 1/16	280.988
9/16	14.2875	1 13/16	46.0376	3 1/2	88.9002	5 3/16	131.763	7 3/4	196.850	11 1/8	282.576
37/64	14.6844	1 27/32	46.8313	3 17/32	89.6939	5 7/32	132.557	7 13/16	198.438	11 3/16	284.163
19/32	15.0812	1 7/8	47.6251	3 9/16	90.4877	5 1/4	133.350	7 7/8	200.025	11 1/4	285.751
39/64	15.4781	1 29/32	48.4188	3 19/32	91.2814	5 9/32	134.144	7 15/16	201.613	11 5/16	287.338
5/8	15.8750	1 15/16	49.2126	3 5/8	92.0752	5 5/16	134.938	8	203.200	11 3/8	288.926
41/64	16.2719	1 31/32	50.0063	3 21/32	92.8689	5 11/32	135.732	8 1/16	204.788	11 7/16	290.513
21/32	16.6687	2	50.8001	3 11/16	93.6627	5 3/8	136.525	8 1/8	206.375	11 1/2	292.101
43/64	17.0656	2 1/32	51.5939	3 23/32	94.4564	5 13/32	137.319	8 3/16	207.963	11 5/16	293.688
11/16	17.4625	2 1/16	52.3876	3 3/4	95.2502	5 7/16	138.113	8 1/4	209.550	11 5/8	295.276
45/64	17.8594	2 3/32	53.1814	3 25/32	96.0439	5 15/32	138.907	8 5/8	211.138	11 11/16	296.863
23/32	18.2562	2 1/8	53.9751	3 13/16	96.8377	5 1/2	139.700	8 3/8	212.725	11 3/4	298.451
47/64	18.6531	2 5/32	54.7688	3 27/32	97.6314	5 17/32	140.494	8 7/16	214.313	11 13/16	300.038
3/4	19.0500	2 3/16	55.5626	3 7/8	98.4252	5 9/16	141.288	8 1/2	215.900	11 7/8	301.626
49/64	19.4469	2 7/32	56.3564	3 29/32	99.2189	5 19/32	142.082	8 9/16	217.488	11 15/16	303.213
25/32	19.8437	2 1/4	57.1501	3 15/16	100.013	5 5/8	142.875	8 5/8	219.075	12	304.801
51/64	20.2406	2 9/32	57.9439	3 31/32	100.806	5 21/32	143.669	8 11/16	220.663		
13/16	20.6375	2 5/16	58.7376	4	101.600	5 11/16	144.463	8 3/4	222.250		
53/64	21.0344	2 11/32	59.5314	4 1/32	102.394	5 23/32	145.257	8 13/16	223.838		
27/32	21.4312	2 3/8	60.3251	4 1/16	103.188	5 3/4	146.050	8 7/8	225.425		

CONVERSION TABLE: DECIMALS OF AN INCH INTO MILLIMETERS

Inches	Milli-meters	Inches	Milli-meters	Inches	Milli-meters	Inches	Milli-meters	Inches	Milli-meters
0.001	0.025	0.140	3.56	0.360	9.14	0.580	14.73	0.800	20.32
0.002	0.051	0.150	3.81	0.370	9.40	0.590	14.99	0.810	20.57
0.003	0.076	0.160	4.06	0.380	9.65	0.600	15.24	0.820	20.83
0.004	0.102	0.170	4.32	0.390	9.91	0.610	15.49	0.830	21.08
0.005	0.127	0.180	4.57	0.400	10.16	0.620	15.75	0.840	21.34
0.006	0.152	0.190	4.83	0.410	10.41	0.630	16.00	0.850	21.59
0.007	0.178	0.200	5.08	0.420	10.67	0.640	16.26	0.860	21.84
0.008	0.203	0.210	5.33	0.430	10.92	0.650	16.51	0.870	22.10
0.009	0.229	0.220	5.59	0.440	11.18	0.660	16.76	0.880	22.35
0.010	0.254	0.230	5.84	0.450	11.43	0.670	17.02	0.890	22.61
0.020	0.508	0.240	6.10	0.460	11.68	0.680	17.27	0.900	22.86
0.030	0.762	0.250	6.35	0.470	11.94	0.690	17.53	0.910	23.11
0.040	1.016	0.260	6.60	0.480	12.19	0.700	17.78	0.920	23.37
0.050	1.270	0.270	6.86	0.490	12.45	0.710	18.03	0.930	23.62
0.060	1.524	0.280	7.11	0.500	12.70	0.720	18.29	0.940	23.88
0.070	1.778	0.290	7.37	0.510	12.95	0.730	18.54	0.950	24.13
0.080	2.032	0.300	7.62	0.520	13.21	0.740	18.80	0.960	24.38
0.090	2.286	0.310	7.87	0.530	13.46	0.750	19.05	0.970	24.64
0.100	2.540	0.320	8.13	0.540	13.72	0.760	19.30	0.980	24.89
0.110	2.794	0.330	8.38	0.550	13.97	0.770	19.56	0.990	25.15
0.120	3.048	0.340	8.64	0.560	14.22	0.780	19.81	1.000	25.40
0.130	3.302	0.350	8.89	0.570	14.48	0.790	20.07		

CONVERSION TABLE: MILLIMETERS INTO INCHES

Milli-meters	Inches	Milli-meters	Inches	Milli-meters	Inches	Milli-meters	Inches	Milli-meters	Inches	Milli-meters	Inches
1	0.0394	33	1.2992	65	2.5590	97	3.8189	129	5.0787	161	6.3386
2	0.0787	34	1.3386	66	2.5984	98	3.8583	130	5.1181	162	6.3779
3	0.1181	35	1.3779	67	2.6378	99	3.8976	131	5.1575	163	6.4173
4	0.1575	36	1.4173	68	2.6772	100	3.9370	132	5.1968	164	6.4567
5	0.1968	37	1.4567	69	2.7165	101	3.9764	133	5.2362	165	6.4960
6	0.2362	38	1.4961	70	2.7559	102	4.0157	134	5.2756	166	6.5354
7	0.2756	39	1.5354	71	2.7953	103	4.0551	135	5.3149	167	6.5748
8	0.3150	40	1.5748	72	2.8346	104	4.0945	136	5.3543	168	6.6142
9	0.3543	41	1.6142	73	2.8740	105	4.1338	137	5.3937	169	6.6535
10	0.3937	42	1.6535	74	2.9134	106	4.1732	138	5.4331	170	6.6929
11	0.4331	43	1.6929	75	2.9527	107	4.2126	139	5.4724	171	6.7323
12	0.4724	44	1.7323	76	2.9921	108	4.2520	140	5.5118	172	6.7716
13	0.5118	45	1.7716	77	3.0315	109	4.2913	141	5.5512	173	6.8110
14	0.5512	46	1.8110	78	3.0709	110	4.3307	142	5.5905	174	6.8504
15	0.5905	47	1.8504	79	3.1102	111	4.3701	143	5.6299	175	6.8897
16	0.6299	48	1.8898	80	3.1496	112	4.4094	144	5.6693	176	6.9291
17	0.6693	49	1.9291	81	3.1890	113	4.4488	145	5.7086	177	6.9685
18	0.7087	50	1.9685	82	3.2283	114	4.4882	146	5.7480	178	7.0079
19	0.7480	51	2.0079	83	3.2677	115	4.5275	147	5.7874	179	7.0472
20	0.7874	52	2.0472	84	3.3071	116	4.5669	148	5.8268	180	7.0866
21	0.8268	53	2.0866	85	3.3464	117	4.6063	149	5.8861	181	7.1260
22	0.8661	54	2.1260	86	3.3858	118	4.6457	150	5.9055	182	7.1653
23	0.9055	55	2.1653	87	3.4252	119	4.6850	151	5.9449	183	7.2047
24	0.9449	56	2.2047	88	3.4646	120	4.7244	152	5.9842	184	7.2441
25	0.9842	57	2.2441	89	3.5039	121	4.7638	153	6.0236	185	7.2834
26	1.0236	58	2.2835	90	3.5433	122	4.8031	154	6.0630	186	7.3228
27	1.0630	59	2.3228	91	3.5827	123	4.8425	155	6.1023	187	7.3622
28	1.1024	60	2.3622	92	3.6220	124	4.8819	156	6.1417	188	7.4016
29	1.1417	61	2.4016	93	3.6614	125	4.9212	157	6.1811	189	7.4409
30	1.1811	62	2.4409	94	3.7008	126	4.9606	158	6.2205	190	7.4803
31	1.2205	63	2.4803	95	3.7401	127	5.0000	159	6.2598	191	7.5197
32	1.2598	64	2.5197	96	3.7795	128	5.0394	160	6.2992	192	7.5590

CONVERSION TABLE: MILLIMETERS INTO INCHES (CONT'D)

Milli-meters	Inches	Milli-meters	Inches	Milli-meters	Inches	Milli-meters	Inches	Milli-meters	Inches	Milli-meters	Inches
193	7.5984	261	10.2756	329	12.9527	397	15.6299	465	18.3070	533	20.9842
194	7.6378	262	10.3149	330	12.9921	398	15.6693	466	18.3464	534	21.0236
195	7.6771	263	10.3543	331	13.0315	399	15.7086	467	18.3858	535	21.0629
196	7.7165	264	10.3937	332	13.0708	400	15.7480	468	18.4252	536	21.1023
197	7.7559	265	10.4330	333	13.1102	401	15.7874	469	18.4645	537	21.1417
198	7.7953	266	10.4724	334	13.1496	402	15.8267	470	18.5039	538	21.1811
199	7.8346	267	10.5118	335	13.1889	403	15.8661	471	18.5433	539	21.2204
200	7.8740	268	10.5512	336	13.2283	404	15.9055	472	18.5826	540	21.2598
201	7.9134	269	10.5905	337	13.2677	405	15.9448	473	18.6220	541	21.2992
202	7.9527	270	10.6299	338	13.3071	406	15.9842	474	18.6614	542	21.3385
203	7.9921	271	10.6693	339	13.3464	407	16.0236	475	18.7007	543	21.3779
204	8.0315	272	10.7086	340	13.3858	408	16.0630	476	18.7401	544	21.4173
205	8.0708	273	10.7480	341	13.4252	409	16.1023	477	18.7795	545	21.4566
206	8.1102	274	10.7874	342	13.4645	410	16.1417	478	18.8189	546	21.4960
207	8.1496	275	10.8267	343	13.5039	411	16.1811	479	18.8582	547	21.5354
208	8.1890	276	10.8661	344	13.5433	412	16.2204	480	18.8976	548	21.5748
209	8.2283	277	10.9055	345	13.5826	413	16.2598	481	18.9370	549	21.6141
210	8.2677	278	10.9449	346	13.6220	414	16.2992	482	18.9763	550	21.6535
211	8.3071	279	10.9842	347	13.6614	415	16.3385	483	19.0157	551	21.6929
212	8.3464	280	11.0236	348	13.7008	416	16.3779	484	19.0551	552	21.7322
213	8.3858	281	11.0630	349	13.7401	417	16.4173	485	19.0944	553	21.7716
214	8.4252	282	11.1023	350	13.7795	418	16.4567	486	19.1338	554	21.8110
215	8.4645	283	11.1417	351	13.8189	419	16.4960	487	19.1732	555	21.8503
216	8.5039	284	11.1811	352	13.8582	420	16.5354	488	19.2126	556	21.8897
217	8.5433	285	11.2204	353	13.8976	421	16.5748	489	19.2519	557	21.9291
218	8.5827	286	11.2598	354	13.9370	422	16.6141	490	19.2913	558	21.9685
219	8.6220	287	11.2992	355	13.9763	423	16.6535	491	19.3307	559	22.0078
220	8.6614	288	11.3386	356	14.0157	424	16.6929	492	19.3700	560	22.0472
221	8.7008	289	11.3779	357	14.0551	425	16.7322	493	19.4094	561	22.0866
222	8.7401	290	11.4173	358	14.0945	426	16.7716	494	19.4488	562	22.1259
223	8.7795	291	11.4567	359	14.1338	427	16.8110	495	19.4881	563	22.1653
224	8.8189	292	11.4960	360	14.1732	428	16.8504	496	19.5275	564	22.2047
225	8.8582	293	11.5354	361	14.2126	429	16.8897	497	19.5669	565	22.2440
226	8.8976	294	11.5748	362	14.2519	430	16.9291	498	19.6063	566	22.2834
227	8.9370	295	11.6141	363	14.2913	431	16.9685	499	19.6456	567	22.3228
228	8.9764	296	11.6335	364	14.3307	432	17.0078	500	19.6850	568	22.3622
229	9.0157	297	11.6929	365	14.3700	433	17.0472	501	19.7244	569	22.4015
230	9.0551	298	11.7323	366	14.4094	434	17.0866	502	19.7637	570	22.4409
231	9.0945	299	11.7716	367	14.4488	435	17.1259	503	19.8031	571	22.4803
232	9.1338	300	11.8110	368	14.4882	436	17.1653	504	19.8425	572	22.5196
233	9.1732	301	11.8504	369	14.5275	437	17.2047	505	19.8818	573	22.5590
234	9.2126	302	11.8897	370	14.5669	438	17.2441	506	19.9212	574	22.5984
235	9.2519	303	11.9291	371	14.6063	439	17.2834	507	19.9606	575	22.6377
236	9.2913	304	11.9685	372	14.6456	440	17.3228	508	20.0000	576	22.6771
237	9.3397	305	12.0078	373	14.6850	441	17.3622	509	20.0393	577	22.7165
238	9.3701	306	12.0472	374	14.7244	442	17.4015	510	20.0787	578	22.7559
239	9.4094	307	12.0866	375	14.7637	443	17.4409	511	20.1181	579	22.7952
240	9.4488	308	12.1260	376	14.8031	444	17.4803	512	20.1574	580	22.8346
241	9.4882	309	12.1653	377	14.8425	445	17.5196	513	20.1968	581	22.8740
242	9.5275	310	12.2047	378	14.8819	446	17.5590	514	20.2362	582	22.9133
243	9.5669	311	12.2441	379	14.9212	447	17.5984	515	20.2755	583	22.9527
244	9.6063	312	12.2834	380	14.9606	448	17.6378	516	20.3149	584	22.9921
245	9.6456	313	12.3228	381	15.0000	449	17.6771	517	20.3543	585	23.0314
246	9.6850	314	12.3622	382	15.0393	450	17.7165	518	20.3937	586	23.0708
247	9.7244	315	12.4015	383	15.0787	451	17.7559	519	20.4330	587	23.1102
248	9.7638	316	12.4409	384	15.1181	452	17.7952	520	20.4724	588	23.1496
249	9.8031	317	12.4803	385	15.1574	453	17.8346	521	20.5118	589	23.1889
250	9.8425	318	12.5197	386	15.1968	454	17.8740	522	20.5511	590	23.2283
251	9.8819	319	12.5590	387	15.2362	455	17.9133	523	20.5905	591	23.2677
252	9.9212	320	12.5984	388	15.2756	456	17.9527	524	20.6299	592	23.3070
253	9.9606	321	12.6378	389	15.3149	457	17.9921	525	20.6692	593	23.3464
254	10.0000	322	12.6771	390	15.3543	458	18.0315	526	20.7086	594	23.3858
255	10.0393	323	12.7165	391	15.3937	459	18.0708	527	20.7480	595	23.4251
256	10.0787	324	12.7559	392	15.4330	460	18.1102	528	20.7874	596	23.4645
257	10.1181	325	12.7952	393	15.4724	461	18.1496	529	20.8267	597	23.5039
258	10.1575	326	12.8346	394	15.5118	462	18.1889	530	20.8661	598	23.5433
259	10.1968	327	12.8740	395	15.5511	463	18.2283	531	20.9055	599	23.5826
260	10.2362	328	12.9134	396	15.5905	464	18.2677	532	20.9448	600	23.6220

CONVERSION TABLE: HUNDREDTHS OF A MILLIMETER INTO INCHES

Milli-meters	Inches	Milli-meters	Inches	Milli-meters	Inches	Milli-meters	Inches	Milli-meters	Inches	Milli-meters	Inches
0.01	0.0004	0.18	0.0071	0.35	0.0138	0.52	0.0205	0.69	0.0272	0.86	0.0339
0.02	0.0008	0.19	0.0075	0.36	0.0142	0.53	0.0209	0.70	0.0276	0.87	0.0343
0.03	0.0012	0.20	0.0079	0.37	0.0146	0.54	0.0213	0.71	0.0280	0.88	0.0346
0.04	0.0016	0.21	0.0083	0.38	0.0150	0.55	0.0217	0.72	0.0283	0.89	0.0350
0.05	0.0020	0.22	0.0087	0.39	0.0154	0.56	0.0220	0.73	0.0287	0.90	0.0354
0.06	0.0024	0.23	0.0091	0.40	0.0157	0.57	0.0224	0.74	0.0291	0.91	0.0358
0.07	0.0028	0.24	0.0094	0.41	0.0161	0.58	0.0228	0.75	0.0295	0.92	0.0362
0.08	0.0031	0.25	0.0098	0.42	0.0165	0.59	0.0232	0.76	0.0299	0.93	0.0366
0.09	0.0035	0.26	0.0102	0.43	0.0169	0.60	0.0236	0.77	0.0303	0.94	0.0370
0.10	0.0039	0.27	0.0106	0.44	0.0173	0.61	0.0240	0.78	0.0307	0.95	0.0374
0.11	0.0043	0.28	0.0110	0.45	0.0177	0.62	0.0244	0.79	0.0311	0.96	0.0378
0.12	0.0047	0.29	0.0114	0.46	0.0181	0.63	0.0248	0.80	0.0315	0.97	0.0382
0.13	0.0051	0.30	0.0118	0.47	0.0185	0.64	0.0252	0.81	0.0319	0.98	0.0386
0.14	0.0055	0.31	0.0122	0.48	0.0189	0.65	0.0256	0.82	0.0323	0.99	0.0390
0.15	0.0059	0.32	0.0126	0.49	0.0193	0.66	0.0260	0.83	0.0327	1.00	0.0394
0.16	0.0063	0.33	0.0130	0.50	0.0197	0.67	0.0264	0.84	0.0331		
0.17	0.0067	0.34	0.0134	0.51	0.0201	0.68	0.0268	0.85	0.0335		

METRIC PREFIXES

micro	= 10^{-6}	(example)	micron
milli	= 10^{-3}		millimeter
centi	= 10^{-2}		centimeter
deci	= 10^{-1}		decimeter
deka	= 10		decaliter
hecto	= 10^{2}		hectoliter
kilo	= 10^{3}		kilometer
mega	= 10^{6}		megaton

LENGTH

1 inch (in.)	= 2.54 centimeters	= 25400 microns
1 foot (ft)	= 12 inches	= .3048 meter
1 yard (yd)	= 3 feet	= .9144 meter
1 mile	= 5280 feet	= 1.609 kilometers
1 nautical mile	= 6080 feet	= 1.853 kilometers
1 millimeter (mm)	= 1000 microns	= .0394 inch
1 centimeter (cm)	= 10 millimeters	= .3937 inch
1 decimeter	= 10 centimeters	= 3.937 inches
1 meter (m)	= 100 centimeters	= 3.281 feet
1 kilometer	= 1000 meters	= .6214 mile

AREA

1 sq in.	=	= 6.45 sq cm
1 sq ft	= 144 sq in.	= .0929 sq meter
1 sq. yard	= 9 sq ft	= .836 sq meter
1 acre	= 43560 sq ft	= .4047 hectare
1 sq mile	= 640 acres	= 259 hectares
1 sq cm	= 100 sq mm	= .155 sq in.
1 sq. meter	= 10,000 sq cm	= 10.764 sq ft
1 hectare	= 10,000 sq meters	= 2.471 acres
1 sq km	= 100 hectares	= .3861 sq mile

WEIGHT, AVOIRDUPOIS

1 ounce (oz)	= 473.5 grains	= 28.35 grams
1 pound (lb)	= 16 ounces	= 453.59 grams
1 pound	= 7000 grains	= .454 kilograms
1 short ton	= 2000 pounds	= 907 kilograms
1 cu ft water @ 4° C.	= 62.42 lb	= 28.31 kilograms
1 gallon water @ 4° C.	= 8.34 lb	= 3.78 kilograms
1 gram (g)	= 1000 milligrams	= 15.43 grains
1 kilogram (kg)	= 1000 grams	= 2.205 pounds
1 metric ton	= 1000 kilograms	= 2204.6 pounds
1 cu cm water @ 4° C.	= 1 gram	= .035 ounces
1 liter water @ 4° C.	= 1 kilogram	= 2.205 pounds

VOLUME, DRY

1 cu in.	=	=	16.39 cu cm
1 cu ft	= 1728 cu in.	=	.0238 cu meter
1 cu yard	= 27 cu ft	=	.7646 cu meter
1 quart, U.S.	= .0389 cu ft	=	1101 cu cm
1 gallon, U.S.	= 4 quarts	=	4.405 cu decimeters
1 peck	= 2 gallons	=	8.810 cu decimeters
1 bushel, U.S.	= 4 pecks	=	35.239 cu decimeters
1 bushel, U.S.	= 1.244 cu ft		
1 bushel, Imperial	= 1.032 U.S. bushels		
1 cord	= 128 cu ft		
1 cu cm	=	=	.061 cu in.
1 cu. decimeter	= 1000 cu cm	=	61.02 cu in.
1 cu meter	= 1000 cu decimeters	=	1.308 cu yd
1 cu meter		=	35.314 cu ft

VOLUME, LIQUID

1 pint	= 16 fluid ounces	=	.473 liters
1 quart	= 2 pints	=	.946 liter
1 gallon, U.S.	= 4 quarts	=	3.785 liters
1 gallon, U.S.	= 231 cu in.	=	3.785 cu decimeters
1 cu ft	= 7.48 gallon, U.S.	=	28.32 cu decimeters
1 gallon, Imperial	= 1.201 gallon, U.S.	=	
1 liter	= 1000 cu cm	=	1.057 quarts
1 liter	= 1 cu decimeter	=	.0353 cu ft
1 decaliter	= 10 liters	=	2.64 gallon, U.S.
1 cu meter	= 1000 liters	=	264.18 gallon, U.S.

DENSITY

1 lb/cu in.	= 1728 lb/cu ft	=	27.68 gram/cu cm
1 lb/cu ft	= 27 lb/cu yd	=	16.018 kg/cu m
1 gram/cu cm	= 1000 kg/cu m	=	62.43 lb/cu ft

PRESSURE

1 lb/sq in.	= 144 lb/sq ft	=	.0703 kg/sq cm
1 lb/sq in.	= 2.036 in. Hg	=	2.307 ft water @ 4° C.
1 kg/sq cm	= 735.51 mm Hg	=	14.22 lb/sq in.
1 kg/sq cm	= 10 m water @ 4° C.	=	.968 standard atmospheres
1 kg/sq cm	= 1 at		
1 in. Hg	= .491 lb/sq in.	=	1.133 ft water at 4° C.
1 in. water	= 5.20 lb/sq ft	=	.0361 lb/sq in.
1 ft water @ 4° C.	= 62.43 lb/sq ft	=	.0305 kg/sq cm
1 ft water @ 4° C.	= .433 lb/sq in.	=	.883 in. Hg
1 standard atmosphere (Atm)	= 14.7 lb/sq in.	=	29.92 in. Hg
1 standard atmosphere	= 33.9 ft water @ 4° C.	=	760 mm Hg
1 standard atmosphere	= 1.033 kg/sq cm	=	1.0133 bars
1 cm Hg	= .0136 kg/sq cm	=	.1934 lb/sq in.
1 Bar	= 750 mm Hg	=	14.5 lb/sq in.
1 ata	= 1 kg/sq cm absolute		

VELOCITY

1 ft/sec	— .682 miles/hr	— .3048 m/sec
1 mile/hr	— 1.467 ft/sec	— .447 m/sec
1 mile/hr	— .868 knots	— 1.609 km/hr
1 m/sec	— 3.6 km/hr	— 3.28 ft/sec
1 km/hr	— .2778 m/sec	— .621 miles/hr
1 knot	— 1.152 miles/hr	— 1 nautical mile/hr

HEAT, ENERGY, WORK

1 ft lb	— .001285 BTU	— 0.13826 kg-meter
1 joule	— 1 watt - second	— .000948 BTU
1 BTU	— 778.1 ft lb	— .252 kcal
1 KCAL	— 3.968 BTU	— 1000 cal
1 hp-hr	— .746 kw-hr	— 2544.7 BTU
1 kw-hr	— 1.341 hp-hr	— 3413 BTU
1 boiler horsepower	— 33479 BTU/hr	— Evaporation of 34.5 water/hr at 212° F.

SOLID AND LIQUID EXPENDABLE REFRIGERANTS

Evaporating Temperature of dry ice (solid CO_2) at 1 atmosphere	— -109° F.
Heat of sublimation of dry ice at -109° F.	— 246.3 BTU/lb
Specific heat of CO_2 gas	— .2 BTU/lb/°F.
Refrigerating effect of solid CO_2 to gas at 32° F. (246.3 + .2 [109 + 32])	— 274.5 BTU/lb
Evaporating Temperature of liquid carbon dioxide (CO_2) at 1 atmosphere	— -70° F.
Heat of vaporization of liquid CO_2 at -70° F.	— 149.7 BTU/lb
Specific heat of CO_2 gas	— .2 BTU/lb/°F.
Refrigerating effect of liquid CO_2 to gas at 32° F. (149.7 + .2 [70 + 32])	— 170.1 BTU/lb
Evaporating Temperature of liquid nitrogen (N_2) at 1 atmosphere	— -320° F.
Heat of vaporization of liquid N_2 at -320° F.	— 85.67 BTU/lb
Specific heat of N_2 gas	— .248 BTU/lb/°F.
Refrigerating effect of liquid nitrogen to gas at 32° F. (85.67 + .248 [320 + 32])	— 172.97 BTU/lb

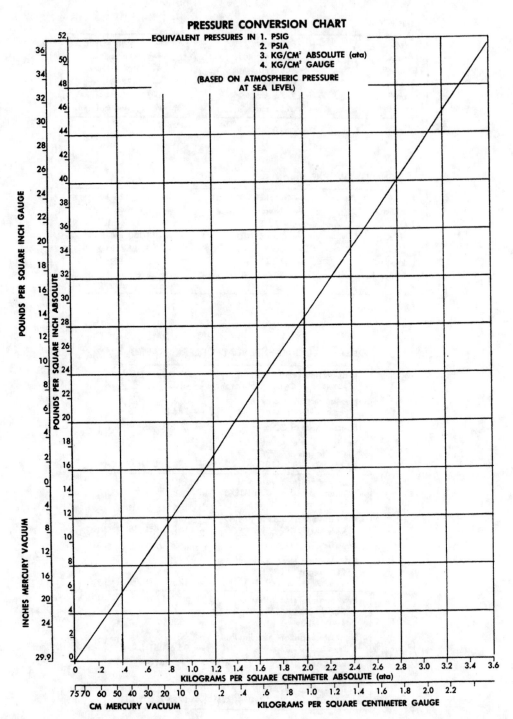

PRESSURE CONVERSION CHART

EQUIVALENT PRESSURES IN 1. PSIG
2. PSIA
3. KG/CM² ABSOLUTE (ata)
4. KG/CM² GAUGE

(BASED ON ATMOSPHERIC PRESSURE
AT SEA LEVEL)

POUNDS PER SQUARE INCH GAUGE

POUNDS PER SQUARE INCH ABSOLUTE

INCHES MERCURY VACUUM

CM MERCURY VACUUM

KILOGRAMS PER SQUARE CENTIMETER ABSOLUTE (ata)

KILOGRAMS PER SQUARE CENTIMETER GAUGE

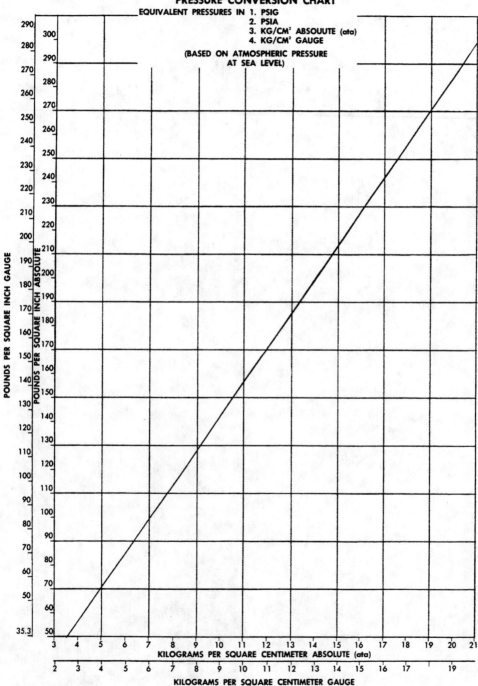

PRESSURE CONVERSION CHART

EQUIVALENT PRESSURES IN 1. PSIG
2. PSIA
3. KG/CM² ABSOUUTE (ata)
4. KG/CM² GAUGE

(BASED ON ATMOSPHERIC PRESSURE
AT SEA LEVEL)

Index

HAVE YOU ORDERED YOUR COPIES OF THESE BILLY C. LANGLEY TITLES?

TO ORDER: Please complete the coupon below and mail to Prentice Hall, Book Distribution Center, Route 59 at Brook Hill Drive, West Nyack, NY 10995.
Or if you prefer, call (201) 767–5937 to place your order.

☐ (R0204–9) AIR CONDITIONING AND REFRIGERATION TROUBLESHOOTING HAND-BOOK, 1980, cloth, $42.67
☐ (R0417–7) BASIC REFRIGERATION, 1982, cloth, $34.00
☐ (R0887–1) COMFORT HEATING, Third Edition, 1985, cloth, $32.00
☐ (17167–8) CONTROL SYSTEMS FOR AIR CONDITIONING AND REFRIGERATION, 1985, cloth, $28.67
☐ (R1036–4) COOLING SYSTEMS TROUBLESHOOTING HANDBOOK, 1986, cloth, $37.33
☐ (24751–0) ELECTRIC CONTROLS FOR REFRIGERATION AND AIR CONDITIONING, Second Edition, 1988, cloth, $30.00
☐ (R1600–7) ELECTRICITY FOR REFRIGERATION AND AIR CONDITIONING, 1984, paper, $27.67
☐ (R1790–6) ESTIMATING AIR CONDITIONING SYSTEMS, 1983, cloth, $32.00
☐ (R2818–4) HEAT PUMP TECHNOLOGY, 1983, cloth, $34.00
☐ (R5578–1) PLANT MAINTENANCE, 1986, cloth, $38.33
☐ (R5638–3) PRINCIPLES AND SERVICE OF AUTOMOTIVE AIR CONDITIONING, 1984, cloth, $32.00
☐ (R6629–1) REFRIGERATION AND AIR CONDITIONING, Third Edition, 1986, cloth, $36.33

Please send me the book(s) checked above. After I've had the chance to examine the book(s) for 15 days, I'll either send the indicated price, plus a small charge for shipping and handling, or return the book(s) and owe nothing. I understand that the publisher will refund the purchase price and permit me to retain, as a complimentary copy, any book that is later ordered in quantities of 10 or more copies by any organization with which I am affiliated.

☐ BILL ME under the terms outlined above.

☐ PAYMENT ENCLOSED—publisher will pay all shipping and handling charges, same full refund, same return privilege as guaranteed above.

NAME/TITLE _____

DEPT/ORGANIZATION _____

ADDRESS _____

CITY _____ STATE _____ ZIP _____

DEPT. 1   D–DSBM–RP(8)